The Trust Algorithm:

How Leaders Build Trust with Generative AI

By Alice Sesay Pope

Printed in the United States of America

This paperback edition: 979-8-9950325-0-2

E-book edition: 979-8-9950325-1-9

Front cover design by DGX NextMedia

Text design and composition by DGX NextMedia

For booking information or bulk sales please email: **contact@alicesesaypope.com**

Alice Sesay Pope

For speaking inquiries, advisory engagements, and board opportunities: contact@alicesesaypope.com
www.alicesesaypope.com

DGX NextMedia
www.dgnxmedia.com

ABOUT THE AUTHOR

Alice Sesay Pope is a Fortune 100 executive who has spent thirty years building customer experience operations that work at scale. Most recently, as Global Vice President of Device, Digital, and Alexa Support at Amazon, she led 10,000 associates across Devices, Alexa, Prime Video, Music and Games serving more than 200 million subscribers across 30+ global sites with over $1 billion in P&L responsibility.

Her career includes senior leadership roles at Amazon, Visa, Capital One, Microsoft, USAA, Johnson & Johnson, and First Horizon. Across those organizations, she has led customer experience transformations in financial services, technology, healthcare, and consumer products, with particular focus on deploying generative AI inside regulated environments where trust is non-negotiable.

She is the author of *Transformation From the Inside Out* (2017) and writes and speaks regularly on the intersection of trust, transformation, product innovation, AI governance, and customer experience at enterprise scale.

Alice serves on the Engineering board at the University of Maryland at College Park and is the founder of Circle of 10, a mentorship program that has shaped a generation of women leaders. She is a limited partner with How Women Invest and holds memberships in the National Association of Corporate Directors, Athena Alliance, and Black Women on Boards.

She lives in the Austin, Texas area with her husband Ed and their sons.

DEDICATION

To Ed — my husband and encourager. You saw this book before I did. You pushed me toward purpose when I needed it most. This book is yours as much as mine.

TABLE OF CONTENTS

INTRODUCTION 9

PART I: THE TRUST CRISIS 14

1. ASSOCIATE, CUSTOMER, AND AI 15

2. THE CUSTOMER EXPERIENCE REVOLUTION 27

3. FROM AI TO GEN AI :THE NEW CX OPERATING SYSTEM 36

4. THE STRATEGIC IMPERATIVE: MARKET REALITY 51

5. GEN AI MYTHS LEADERS MUST LET GO 59

PART II: THE ALGORITHM 67

6. THE EMPATHY + PRECISION EQUATION 68

7. HIGH-IMPACT USE CASES IN CUSTOMER EXPERIENCE 75

8. THE AI ADOPTION J-CURVE MANAGING THE PRODUCTIVITY PARADOX 81

9. THE AGENTIC REVOLUTION HOW AI AUGMENTS WORK WITHOUT REPLACING WORKERS 90

PART III: THE OPERATING SYSTEM 99

10. AI GOVERNANCE 100

11. THE TRANSFORMATION LEADER'S PLAYBOOK 121

12. STRATEGIC FRAMEWORKS 126

13. SCALING TRUST IN A REGULATED WORLD 132

14. THE GEN AI WORKFORCE OF THE FUTURE 139

15. CHANGE MANAGEMENT IN THE GEN AI AGE 144

PART IV: THE SO WHAT 149

16. MEASURING WHAT MATTERS 150

17. LEADERSHIP IN THE AGE OF INTELLIGENT EXPERIENCE 171

18. THE NEXT FRONTIER: AI + HUMANS CREATING VALUE 176

19. INDUSTRY-SPECIFIC TRANSFORMATION INSIGHTS 183

20. THE PATH FORWARD 191

21. THE FUTURE BELONGS TO LEADERS WHO CARE 198

PART V: APPENDIX 213

KEY STATISTICS AND RESEARCH SOURCES 214

INTRODUCTION

We are in a trust recession.

Customers don't trust that their bank will resolve issues without multiple handoffs. Employees don't trust that "transformation" will improve their work rather than eliminate it. Boards don't trust that AI will create advantage without creating reputational, regulatory, or security risk. The signals are everywhere: declining confidence in corporations, persistently low employee engagement, and a stubbornly high failure rate for digital transformation. We have more data, more automation, and more AI capability than at any time in history—yet many organizations are experiencing a widening credibility gap between what they promise and what people experience.

The trust recession was built through thousands of micro-decisions: optimizing efficiency over accountability, automation over ownership, speed over clarity, and cost reduction over customer outcomes. Trust eroded one interaction at a time—every time a customer repeated their story, every time an employee learned about a major change through a broadcast memo, every time policy protected the company at the expense of the customer. And then generative AI arrived.

To many leaders, Gen AI looked like the long-awaited unlock: reduce cost-to-serve, improve productivity, scale service, personalize experiences. The promise is real. But here is what's often missed in the rush to deploy: generative AI doesn't just automate work, it exposes the operating system. It makes workflows visible. It reveals policy ambiguity. It surfaces where an organization has been relying on heroic frontline effort to compensate for systemic friction. AI is not simply a lever for efficiency; it is a spotlight on whether an enterprise deserves trust at scale.

Organizations that deploy AI on top of unresolved root causes aren't transforming—they're accelerating failure.

The leaders I work alongside—COOs, CCOs, Chief Transformation Officers, CIOs—are operating under a mandate that can feel contradictory: deploy AI for efficiency while strengthening trust. Reduce expense while improving experience. Move with urgency while satisfying regulatory expectations. This is the leadership paradox of our time. It is also why I wrote this book.

The Trust Algorithm is not a book about models, prompts, or platforms. It is a book about executive discipline—how to deploy generative AI in a way that compounds trust instead of eroding it. The organizations that will win in the AI era will not simply be those with the largest budgets or the newest tools. They will be the ones with operating systems that deserve belief: systems built on clarity, accountability, empathy, and measurable outcomes.

The leaders who get this right won't just survive the AI revolution; they will define it. They will build organizations where AI elevates human potential instead of diminishing it, where efficiency and empathy compound instead of competing, and where customers and employees both experience systems that deserve their trust.

The question isn't whether AI will transform your business—it will. The question is whether you'll lead that transformation in a way that builds trust or erodes it. This book will show you how to choose wisely.

How This Book Is Organized

This book is structured in four parts, each building on the last that moves from diagnosis to framework to execution to impact.

Part I: The Trust Crisis (Chapters 1–5)

We begin where most leaders are standing right now: at the intersection of rising AI capability and declining customer and employee trust. Chapter 1 reframes the three-way relationship between associates, customers, and AI establishing that trust cannot be engineered without understanding all three. Chapters 2 and 3 trace the evolution of customer experience from transactional to transformational and explain why generative AI has become the new CX operating system. Chapter 4 confronts the market reality of 2025–2026 head-on: the competitive pressures, regulatory headwinds, and board-level expectations that make this moment unlike any before it. Chapter 5 clears the air, naming the Gen AI myths that are causing leaders to stall, over correct, or deploy for the wrong reasons.

Part II: The Algorithm (Chapters 6–9)

This section introduces the core intellectual framework of the book. Chapter 6 presents the Empathy + Precision Equation, the insight that sustainable AI-driven experiences require for both emotional intelligence and operational accuracy. Chapter 7 moves from theory to practice, examining the highest-impact Gen AI use cases in customer experience with real-world applications. Chapter 8 prepares leaders for what happens after deployment: the AI Adoption J-Curve, where productivity dips before it rises, and how to lead through the paradox without losing organizational confidence. Chapter 9 introduces the Agentic Revolution — how AI is evolving from a tool that responds to one that acts, and what that means for how work gets done.

Part III: The Operating System (Chapters 10–15)

Transformation without governance is a liability. Part III is the executive playbook: the decisions, structures, and disciplines leaders must put in place to scale AI responsibly. Chapter 10 addresses AI Governance directly: the ethical, regulatory, and operational frameworks that protect organizations from accelerating the wrong outcomes. Chapter 11 is written specifically for COO's and CCO's who bear the operational weight of this work. Chapter 12 draws on strategic frameworks from McKinsey, BCG, and Harvard Business Review to ground the book's recommendations in the best available external research. Chapter 13 tackles the unique challenge of scaling trust in regulated industries, where speed and compliance must coexist. Chapters 14 and 15 turn to people: the Gen AI workforce of the future, and the change management discipline required to bring employees through transformation.

Part IV: The So What (Chapters 16–21)

The final section answers the question every leader eventually asks: *How do I know if it's working?* Chapter 16 introduces the Trust Scorecard, a practical measurement framework for quantifying what has historically been called "soft." Chapter 17 examines what leadership itself must look like in the age of intelligent experience. Chapter 18 explores the emerging frontier of AI and humans co-creating value and not competing for it. Chapter 19 provides industry-specific transformation insights for 2025–2026, acknowledging that the path forward looks different in financial services, healthcare, retail, and beyond. Chapter 20 maps the road ahead into 2026 and the decisions that will separate leaders from laggards. Chapter 21 closes with a conviction that runs through every page of this book: *the future belongs to leaders who care* — about their customers, their employees, and the kind of organizations they are building.

Whether you read this book cover to cover or navigate directly to the chapter most relevant to your immediate challenge, each section is designed to be actionable. Where relevant, you'll find references to the **Trust Algorithm Diagnostic** — a scorecard tool that allows you to benchmark your organization's trust posture and identify the highest-leverage areas for intervention.

This book will not give you a shortcut. But it will give you a framework, a diagnostic, and a leadership discipline that makes the path forward clear — and the outcome worth the effort.

PART I: THE TRUST CRISIS

1. ASSOCIATE, CUSTOMER, AND AI

The Same Wrong Answer Twice.

She had a straightforward question. The kind of question that should take ninety seconds to answer. The kind of question that exists somewhere between routine and important, the kind customers assume any competent company can handle without incident. It involved her account, a recent policy, and a decision she needed to make before the end of the week.

She did what most customers do. She started with self-service. She opened the app, found the AI assistant, and typed her question in plain language. The assistant responded immediately. It cited a policy and it explained the terms. It told her exactly what she was eligible for and what steps to take next.

She was satisfied. She made her decision based on that answer and moved on with her day. Two days later, something did not add up. A confirmation she expected never arrived. A number on her account did not match what she had been told. A small thread of doubt pulled at the edges of her confidence. So, she did what customers do when self-service leaves them uncertain: she called.

The associate who answered was professional, attentive, and genuinely trying to help. He listened to her question. He pulled up her account. He consulted the knowledge base, the same system the company had built to ensure consistency across every channel, the repository that both AI and associates relied on as the single source of truth.

He gave her the same answer.

Not a similar answer or a version of the answer, but the same answer. The same policy, the same terms, the same confident, specific, and incorrect information that the AI assistant had provided forty-eight hours earlier.

Now, she had confirmation. Two separate interactions, one with a machine, one with a human and told her the same thing. In the customer's mind, this was no longer a data point, it was a fact. Two independent sources had validated the same information. What reason would she have to doubt it?

She acted accordingly. She made a financial decision. She adjusted her plans. She proceeded with full confidence that the company she trusted had given her reliable information through two different channels.

She found the truth herself.

It might have been a friend who had a different experience. It might have been a line in the fine print she finally read closely. It might have been a third interaction with a different agent, or a different part of the company's website where the correct policy was still visible, untouched by whatever had gone wrong in the knowledge base. However she found it, the realization was immediate and absolute: what she had been told twice by two different sources, and which she had every reason to trust, was wrong.

The policy the AI cited did not exist as described. The terms the associate confirmed were outdated. Somewhere in the system, a knowledge article had not been updated, or had been updated incorrectly, or conflicted with another source that carried different information. The root cause, when it was eventually traced, was not dramatic. It was the kind of quiet, unglamorous failure that happens inside large organizations every day: a knowledge base that had not kept pace with a policy change. A source of truth that was no longer true.

The customer did not experience a knowledge management gap. The customer experienced a betrayal. She called back and was furious.

This was the fury of someone who had trusted a system, acted on that trust, been harmed by it, and then discovered on her own that nobody at the company had caught the error or corrected it. She had done the company's job for them. She found the truth: their AI and their people had both failed to provide it.

There is frustration and sometimes anger a customer feels when they realize they know more about your policies than your own systems do. It is not just frustration; it is a loss of faith. It is the moment when a customer stops assuming competence and starts assuming incompetence, when the burden of proof shifts permanently, and every future interaction begins from a position of suspicion rather than trust.

I have led customer experience operations long enough to know that this moment is an expensive one in the customer life cycle, not because of the remediation cost or escalation time, but because of the trust that evaporates and the extraordinary effort required to rebuild it, if it can be rebuilt at all.

The failure was not AI. The failure was not the associate. The failure was the system. This story matters as the opening of this book because it dismantles the two most common assumptions leaders carry into every AI conversation.

The first assumption is that AI is the risk. When executives worry about deploying AI in customer-facing environments, the fear is almost always about technology: What if the bot hallucinates? What if it gives the wrong answer? What if it says something inappropriate? Those are legitimate concerns. But this story reveals something more uncomfortable: the AI gave a wrong answer, not because it hallucinated, not because the model failed, but because it was anchored to a faulty knowledge source. The technology worked exactly as designed. It retrieved information from the authorized source and delivered it with precision. The problem was upstream.

The second assumption is that humans are the safety net. This is the comforting belief that undergirds most AI deployment strategies: even if the bot gets it wrong, the agent will catch it. The human in the loop is the backstop. The escalation path is protection. But in this case, the agent confirmed the AI's wrong answer because the agent was consulting the same corrupt source. The safety net had the same hole.

When both the AI and human draw from the same knowledge base, neither provides independent validation. They provide one validation, delivered twice, with the illusion of independence. Two sources agree, so it must be right. In reality, a single point of failure has simply been amplified across two channels.

This is the system problem that most organizations have not yet confronted. It is not enough to build a good bot. It is not enough to train good associates. If the knowledge infrastructure underneath them is stale, inconsistent, or wrong, every channel you operate becomes a delivery mechanism for the same error. The more channels you have — AI assistant, live chat, phone, email, help center, the more efficiently you distribute bad information to the people who trust you most.

What This Means for the Three-Way Relationship

This book is built on a foundational premise: Associate, Customer, and AI are not three separate priorities. They are one system. This story is why.

The customer in this story did not interact with "an AI" and then "an agent." She interacted with a company. In her mind, there was one entity, one brand, one promise, one relationship. The AI was not a separate product. The agent was not a separate experience. They were two expressions of the same organization, and she expected them to be aligned, accurate, and trustworthy. When they both failed, the failure was not attributed to either channel individually. It belonged to the entire system.

This is the lens I want you to carry through every chapter of this book. When we talk about deploying AI in customer experience, we are not adding a new tool to an existing operation. We are redesigning the system that determines whether trust is built or broken. The AI's performance depends on the knowledge it is given, which depends on the processes that maintain that knowledge, which depend on the people and governance structures responsible for accuracy. The associate's performance depends on the same chain. If any link fails, both fail and the customer pays the price.

The customer does not see your organization chart or channel strategy. They do not distinguish between your AI team, your training team, your knowledge management team, and your quality assurance team. They see a company. They judge the company by the worst answer received, especially when that answer was the same one, given twice.

The Questions This Story Should Raise

Before we go further into the frameworks, strategies, and operating models in the chapters ahead, sit with these questions. They are not rhetorical. They are diagnostic and your honest answers will determine how much value you extract from the rest of this book.

How current is your knowledge base today? Not how current it was at the last audit. How current is it at this moment? If a policy changed last week, does your AI assistant know? Does your agent know? Do they know the same thing? Most leaders cannot answer this question with confidence, and that gap is where stories are like the one in the opening chapter of this book.

Who owns knowledge accuracy in your organization? Not who "contributes" to the knowledge base. Who is accountable with their name, their performance review, their professional reputation attached for ensuring that what your AI says and what your associates say is correct? In many organizations, knowledge is everyone's responsibility and no one's accountability. That is not a governance model. It is an accident waiting to happen.

If your AI and your agent give the same wrong answer today, how long would it remain undetected? Hours? Days? Weeks? Do you even have a mechanism designed to detect this specific failure mode where the error is not a system crash or a hallucination, but a confidently delivered piece of outdated information? Most quality monitoring is built to catch outliers: the associate who goes off script, the bot that generates something bizarre. It is not built to catch the far more dangerous scenario where everyone is on-script and the script is wrong.

When a customer discovers an error you miss, what happens? Is there a rapid feedback loop that corrects the source and notifies every channel or does the correction happen locally? One associate updates their notes; one ticket is resolved while the bad information continues to flow to the next hundred customers.

Are your AI and associates truly independent sources of information, or are they one source wearing two masks? This is the question that cuts to the heart of the matter. If both channels pull from the same knowledge base without independent verification layers, you do not have redundancy. You have a single point of failure that appears to be redundant, which is arguably worse because it creates false confidence in both the customer and the organization.

Why This Story Opens the Book

I chose to start here not with a technology overview, not with market data, not with a framework because this story contains the entire book's thesis in a single customer interaction.

Trust is not built by AI. Trust is not built by associates. Trust is built by the system that connects them with the knowledge that feeds them, the governance that monitors them, the processes that keep them accurate, and the leadership that holds all of it accountable.

When that system works, something remarkable happens. The AI provides fast, accurate, empathetic self-service. The associate provides deep, nuanced, human support for the moments that require it. The customer moves between the two seamlessly, with growing confidence that this is a company that knows what it is doing and has earned their trust. Each interaction reinforces the last. Trust compounds.

When that system breaks even in a small, quiet, undramatic way, like a knowledge article that was not updated, the compounding works in reverse. One wrong answer becomes two. Two wrong answers lead to a customer decision based on false information. A decision made on false information becomes fury. Fury in the age of social media and online reviews and letters to the CEO, does not stay contained.

The Trust Algorithm is the operating system that ensures compounding goes in the right direction. Every variable—Trust Signals, Empathy Engine, Precision Layer, Recovery System, Risk Governance, Learning Loop exists to prevent stories like this one, or to catch them in hours instead of days, or to recover from them in ways that rebuild trust rather than erode it.

In the chapters ahead, we will build each layer of that system. I want you to carry thoughts of this customer with you through the book but as an abstraction. As a person who asked a simple question, trusted the answer, trusted it again when a human confirmed it, and then discovered it on her own that the company she relied on had failed her twice.

She is every customer. She is your customer. The system that failed her is the system this book will teach you to fix.

Associate. Customer. AI. The three are not separate priorities. They are one system. The system is only as trustworthy as its weakest source of truth.

When we design AI to elevate the associate, we create faster, more human customer experiences and we build operations that scale without losing the soul of the brand. The pages ahead translate that belief into practical, measurable use cases that improve customer experience, reduce cost-to-serve, and strengthen trust at the moments that matter most.

Consider what this looks like in the real world. A customer calls angry after multiple failed password resets. Instead of forcing the associate to hunt through scattered knowledge bases and past transcripts, a Gen AI copilot instantly summarizes the customer's history, surfaces the right troubleshooting flow, and prompts the associate to acknowledge the customer's frustration before acting.

In another scenario, a subscription business sees the same billing defect trigger repeat contacts week after week. Gen AI detects the pattern at-scale, clusters the most common drivers, and produces a clear defect brief for Product, accelerating root-cause fixes that can reduce repeat contacts in days, not quarters. Picture a frontline leader responsible for coaching quality across a distributed workforce. With Gen AI-derived signals (tone, resolution indicators, policy adherence, and friction points), quality assurance can move from sampling to broad coverage, enabling targeted coaching that improves first-contact resolution without creating unhealthy pressure to minimize handle time.

Imagine a single interaction that brings all of this to life. A customer calls, frustrated that they purchased a product that isn't working properly after spending hours trying to fix it themselves. The tone carries exhaustion and skepticism: "Here we go again, another call, another script, another waste of time." In many organizations, that moment becomes predictable. The associate follows a rigid sequence, the customer feels unseen, and the interaction ends with either escalation or resignation.

Even when a Gen AI-enabled operation is designed correctly, something can shift within seconds. The associate is guided in real time by a coaching layer that provides context from prior interactions, recommends empathetic phrasing, and accelerates resolution. The associate is no longer buried in manuals or toggling between systems; they are free to focus on the human being on the other end. The outcome is not simply a faster fix. The outcome is a different emotional trajectory: from frustration to relief, from skepticism to confidence.

By the end of that interaction, the customer is laughing. They do not thank the algorithm. They thank the person. The most durable advantage in modern business is not automation. It is trust created by people, supported by systems, and reinforced through consistent experiences.

From Cost Center to Growth Engine

Leadership attention was often concentrated on operational efficiency: average handle time, cost per contact, staffing ratios, and utilization. Those metrics matter, but they are incomplete. They describe the cost of interaction, not the value created (or destroyed) by that interaction.

The landscape has changed. Customer experience is now a board-level topic because it directly shapes revenue performance, brand resilience, and long-term enterprise value. The switching costs for customers have collapsed; in many categories, a single bad experience can send a customer to a competitor instantly. On the other hand, a single exceptional experience can create years of loyalty, repeat purchase behavior, and referral momentum. Across industries from retail to financial services to healthcare, trust has become a growth engine, and customer operations are among the most leveraged places to build it.

Why Gen AI Is Different

Traditional AI brought meaningful efficiencies. Predictive models improved forecasting and routing. Rules-based automation reduced manual work. Early bots handled basic transactions, but they struggled with nuance, tone, context, and the messy reality of human emotion. It could process inputs, but it rarely created a feeling of being understood.

Generative AI changes the shape of the problem. It can communicate in natural language, adapt responses in real time, and turn fragmented operational knowledge into immediate guidance for the frontline. When implemented responsibly, Gen AI reduces the cognitive load on associates, shortens time to resolution, and raises the standard of care without sacrificing speed. The important shift is not that Gen AI acts faster; it enables better decisions at the edge of customer experience, where trust is either earned or lost.

This is not technology for technology's sake. It is an opportunity to unlock a new layer of humanity in customer interactions at-scale by giving people better tools, clearer context, and stronger support in the moments that define the relationship.

Leaders at the Crossroads

Every leadership team now faces a strategic choice. Will Gen AI be treated as a blunt cost-cutting instrument—focused primarily on headcount reduction and containment? Will it be deployed as a strategic enabler or a trust amplifier that elevates employees, strengthens customer relationships, and fuels sustainable growth?

History offers a cautionary pattern. Organizations that pursue efficiency without empathy often optimize processes while quietly eroding loyalty. Customers begin to feel like transactions. Employees feel like interchangeable parts. The brand loses the very differentiation it needs to win in an experience-driven market. The future will belong to leaders who pursue a different equation: efficiency with humanity, scale with trust, automation with accountability.

The Empathy Imperative

Customers do not only want speed. They want respect for their time, clarity in communication, and evidence that the company understands what they are going through. Empathy has become a measurable business capability, and Gen AI creates a path to scale it.

A Leadership Test

Across my career leading global operations at Amazon, Microsoft, Visa, USAA, and Capital One, I have learned that every wave of technology is also a test of leadership. Outsourcing reduced costs, but organizations that paired it with strong quality systems gained loyalty.

Digital self-service improved convenience, but organizations that paired it with empathy and seamless escalation created trust. Gen AI will deliver efficiency for many companies but the leaders who harness it to reimagine customer experience, elevate associates, and create measurable trust will define the next decade of competitive advantage.

A customer's smile is not powered by code alone. It is powered by a human equipped with the right tools, supported by intelligent systems, and empowered to show up with clarity and care. That is where this journey begins: with a question every leader must answer honestly.

Are you using AI to only cut costs or are you using it to create delightful experiences, lasting loyalty and increased revenue?

2. THE CUSTOMER EXPERIENCE REVOLUTION

The shift that changed everything. Consider a different version of the story I told in Chapter 1.

A customer notices an unexpected charge on her account; one she does not recognize. She opens the app and types a question into the AI assistant. The assistant not only answers her questions but also provides additional information. It recognizes the pattern: this specific charge is associated with a recent policy update that has generated a spike in similar inquiries over the past seventy-two hours. The knowledge base was updated within hours of the policy change. The AI pulls from that current source and responds with language that is specific, accurate, and written in plain terms: here is what the charge is, here is why it appeared, and here is exactly what will happen next.

The customer feels a small wave of relief, but the system does not stop there. Because the AI bot flagged this as a high-confusion issue, it proactively offers a one-click option to connect with a live associate if she wants further reassurance. She does not need it, the answer was clear, but the option itself communicates something: we know this might feel unsettling, and we are here if you need us.

Three days later, she receives a brief follow-up message confirming the charge was processed as expected and offering a link to updated policy details. She did not ask for this. The system anticipated that customers who inquired about this charge might want confirmation once it is settled.

The customer never calls. She never has questions about whether the company has her back—because every signal she received, across every channel, at every moment, told the same accurate, empathetic, and consistent story.

That is what a system looks like when Associate, Customer, and AI are operating as one. In Chapter 1, the customer got the same wrong answer twice because the knowledge base was broken, and no one caught it. In this version, the customer got the right answer once and never needed to ask again because the knowledge was up to date, the AI was anchored, and the system was designed for confidence rather than deflection. The difference is not technology. It is the operating discipline behind it.

Today, customer experience is business. It's the strategy customers feel, the culture employees live, and the operating system that determines whether trust compounds or collapses interaction by interaction. Customer experience (CX) is no longer simply about satisfaction. It is about confidence. It is about whether customers believe you will deliver on what you promised, whether they can succeed without friction, and whether you will take ownership when things go wrong. Those outcomes are not "soft." They show up in churn, retention, cost-to-serve, brand preference, and the speed at which your business can grow without breaking.

Customers no longer compare you to your direct competitors. They compare you to the best experience they had anywhere. The best bank app they used. The best delivery experience they had. The most intuitive streaming interface. The easiest returns journey. The most respectful support conversation. Expectations have become portable, and customers carry them from industry to industry. That is the revolution: your benchmark is not the company across town; it's the best interaction your customer had this morning.

Why CX Is the New Growth Engine

For years, companies chased growth primarily through more: more marketing spend, more features, more channels, more campaigns, more segmentation. In a world where switching is easy, and patience is thin, growth is increasingly decided by one question: How easy do you make it for customers to succeed?

Modern CX is not "how we make customers happy." It is how we reduce friction, build confidence, and earn repeat business. When experience is engineered as a system measured, governed, and improved with the same rigor as a revenue line, CX becomes a growth engine that produces outcomes executives can take to the board. Retention that is durable, and advocacy that lowers acquisition costs, improves margins.

When CX is done well, customers stay because they trust you. They buy more because it's easier. They recommend your company because you delivered. The highest-performing companies do not treat experience as subjective sentiment; they treat it as a measurable system that produces repeatable results.

Forces Powering the Revolution

This shift didn't happen because companies suddenly became kinder. It happened because forces collided—and together they rewired the economic logic of customer operations. First, digital habits permanently reshaped expectations. Customers now assume instant access, personalization, transparency, self-service that works, and fast recovery when something breaks. "Hold please" isn't merely inconvenient; it feels outdated, like a business hasn't kept pace with the world its customers live in. Customers don't want to learn your processes. They want your processes to respect their time.

Second, trust became the currency. Customers are deciding consciously and subconsciously whether they can trust you with their time, money, data, and peace of mind. Trust isn't built through slogans or brand messaging. It is built through consistency: the reliability of your product, the clarity of your communication, and your willingness to take ownership when outcomes fall short. At the same time, employees are asking their own trust questions: Will leadership be fair? Will my role evolve or disappear? Will I be trained and supported or replaced? The customer and employee trust equations are now inseparable. You cannot build durable customer trust on top of internal uncertainty and fear.

This is where the leadership mandate becomes unavoidable: after you improve automation, are you also taking care of your employees? Do they trust the transformation? Do they have the skills, tools, and clarity to succeed? The best Gen AI strategy fails if it is not paired with a trust strategy, internally and externally.

From Moments to Systems

Many organizations chase "wow moments." They rewrite scripts, launch empathy training, introduce a new survey, deploy a chatbot, and declare transformation. World-class CX leaders think in systems because they understand a simple truth: service is often the symptom, not the disease. Sustainable improvement comes from managing the connected mechanisms that create end-to-end experiences. In practice, that means leaders build and govern five interdependent systems.

The first is the **Experience System**: what customers are trying to accomplish, where they struggle, and how easily they can complete the journey across channels. This system forces clarity about the customer's job-to-be-done and where friction is being created.

The second is the **Defect System**: what repeatedly breaks, why it persists, and how quickly the organization detects and eliminates root causes. This is where "contacts" become intelligence signals of what the enterprise must fix upstream.

The third is the **People System**: hiring, training, tools, empowerment, coaching, and change management. This system determines whether the frontline is confident, capable, and supported or overwhelmed and reactive.

The fourth is the **Knowledge System**: what the organization knows about customers, what associates know in the moment, and how quickly that knowledge stays current as products and policies change. In the world of Gen AI, knowledge becomes a living asset; when it's wrong, everything downstream becomes expensive.

The fifth is the **Measurement System**: metrics that tell the truth about effort, resolution, trust, and value not just speed and volume. This is where leaders decide what the organization will optimize for, and what tradeoffs will be tolerated.

When these systems align, the company doesn't just "deliver service." It delivers confidence. Confidence drives retention, expansion, and brand resilience.

The New Standard

The modern standard for CX can be summarized in three qualities: effortless, proactive, and human when it matters.

Effortless does not mean customers want to be "taken care of." It means they want the experience to be simple, fast, and correct. Simple means fewer steps and clearer language. Fast means less waiting, less searching, less repeating, and fewer transfers. Correct means the right outcome, delivered on the right channel that makes sense for the customer. The deeper point executives should notice is this: when the customer experience is effortless, customers do less work, employees do less re-working, and the business pays less for the same or better outcome.

Customers don't always need a person, until they do. High-stakes moments demand empathy without scripts, ownership without handoffs, clear options, fast recovery, and dignity. These are the moments where loyalty is decided. If automation is designed to deflect customers during high-stakes events, trust collapses. If automation is designed to accelerate resolution and elevate the human connection when needed, trust compounds.

The Myth That Holds Companies Back

A common myth still blocks progress: "CX is the job of customer service." The reality is that the contact center is often the cleanup crew for decisions made elsewhere, confusing product design, unclear policies, broken fulfillment, billing defects, gaps in self-service, misaligned incentives, and disconnected ownership.

Your support organization cannot outwork a broken upstream experience. If the enterprise continues to push defects into the contact center, the contact center will always look "inefficient," no matter how hard the team works. That is why the most effective CX leaders don't just manage interactions. They lead cross-functional accountability: Product, Engineering, Operations, Marketing, Finance, Legal/Compliance, and vendor partners, because that is where the real levers live.

This is the revolution: CX is now enterprise leadership.

The CX Leadership Model for This Era

Throughout this book, I want you to carry out a simple model that makes CX measurable and operational, not abstract:

Revolutionary CX = Trust + Ease + Recovery + Value (TERV)

Trust is whether customers and employees believe you will do what you said.

Ease is whether customers can succeed without unnecessary effort. Recovery is how quickly and responsibly you take ownership when things go wrong.

Value is whether customers feel the relationship is worth their time, money, and energy.

When leaders improve **TERV**, outcomes follow.

Retention improves because trust increases. Expansion improves because ease and value rise. Cost-to-serve falls because effort and repeat contacts drop. Brand resilience strengthens because recovery becomes a capability, not an improvisation.

Apply TERV to the story in Chapter 1, and the failure becomes surgically clear. Trust failed because the customer received fabricated information from both the AI and the agent. Ease failed because the customer eventually had to do the company's job herself, discovering the correct policy through her own research. Recovery failed because no one caught the error before she did. The only variable that remained intact was **Value**, the customer still believed the product itself was worth having. That residual belief in value is the only reason she called back furious instead of leaving quietly. She was giving the company one more chance.

Most customers do not give you that chance. They simply leave.

Markers That You're Operating in the New Era

You can tell you've entered the CX revolution when customer experience metrics are reviewed with the same rigor as revenue metrics and cost metrics. You can see it when product teams are accountable for defects that drive contact, when self-service is designed for customer success rather than deflection, and when frontline teams have tools and authority not just scripts and escalation paths.

You can also see how AI is used. In the past, technology was often deployed to reduce contact by pushing customers away from humans, regardless of whether the issue was resolved. In current times, AI is deployed to remove friction and increase resolution, with escalation to humans handled intelligently for high-stakes moments. Leaders track repeat contacts and customer effort as seriously as they track cost. They don't ask, "How do we handle more contacts?" They ask, "Why are customers contacting us, and how do we eliminate the cause, while supporting employees to be more effective?"

A Practical Executive Checklist

Here is a practical checklist you can use with any organization, whether you lead a customer function or sit on an executive team responsible for enterprise performance. Start with the basics:

- What are the top reasons customers contact us, and which are preventable?

- Where do customers lose time, repeat themselves, or get transferred across teams?
- Which policies create friction, and who has the authority to change them?
- How quickly do we detect emerging issues before they become floods?
- Is self-service designed for success, or designed for deflection? Do you measure resolution and effort, or mainly speed and volume?
- If a customer has a bad day with you, how quickly can we regain their trust?

Finally, in a Gen AI transformation, how are you ensuring employees are engaged, trained, and aligned, so the change becomes capacity, not chaos? If these questions can be answered with data and action, not anecdotes and intentions, you are leading the revolution.

Closing: This Is Your Advantage

The CX revolution is not a trend. It is a permanent reset of expectations. The companies that win will not be the ones with the most sophisticated or loudest messaging. They will be the ones that build trustworthy, effortless, human-centered systems powered by AI, guided by strong leadership, and anchored in real accountability. Customer experience is not what you intend. It is what customers remember and what they decide to do next.

3. FROM AI TO GEN AI :THE NEW CX OPERATING SYSTEM

The Moment Everything Changed

Traditional AI has been shaping the customer's experience for years often quietly, in the background, and largely out of customers' sight. It optimized routing, detected fraud, powered recommendations, improved forecasting, and executed "if-then" decisioning at-scale. In most organizations, this version of AI made operations faster and more efficient, and in the best cases, it made them smarter. But it rarely touched the most human part of the journey: the conversation itself. It sat outside the moment where trust is formed, where a customer is anxious, confused, angry, or simply trying to get something done without friction.

Then generative AI arrived and shifted the center of gravity.

Not because it is "better automation," and not because it replaces every prior AI technique. Gen AI changes the game by expanding what machines can do in language, interpretation, and decision support directly within the customer interaction and alongside the employee delivering it. The implication for leaders is bigger than a new toolset. It is a structural shift: when systems can interpret messy human intent and generate coherent, context-aware responses, the company's operating model, not just its technology stack must evolve.

This chapter is about that shift: what it means, what it does not mean, and how to turn Gen AI into a durable advantage rather than a shiny distraction. The organizations that win won't be the ones that deploy the most chatbots. They'll be the ones that redesign how knowledge flows, how decisions are made, and how accountability works, so AI strengthens trust instead of quietly eroding it.

You can see the shift in practical applications that create immediate executive-level outcomes. At the frontline, an agent copilot can generate a concise case summary, draft a follow-up email aligned with policy tone, and automatically log notes, reducing after-call work while improving quality consistency across thousands of employees. In governance, Gen AI can continuously audit knowledge articles against current policy, flag contradictions, and surface "source-of-truth drift" so leaders can fix the underlying content before it causes customer harm. In operational resilience, a CX organization can use Gen AI to detect emerging issue patterns such as a new firmware defect within hours rather than weeks, triggering earlier recovery communications and reducing downstream contact volume. These are not "AI demos." These are structural improvements to the systems that shape trust, cost-to-serve, and recovery speed.

AI vs. Gen AI: What's Different

Traditional AI excels when the problem is definable, and the desired outcome is known. It predicts outcomes such as churn risk, demand, and fraud likelihood. It classifies and routes work based on intent detection and structured signals. It optimizes decisions like workforce planning and next-best-action selection. It detects anomalies unusual behavior, spikes, defects when you know what "normal" looks like and you have the right data instrumentation. Traditional AI is powerful, but it typically works best when inputs are structured, and the outputs can be evaluated against a clear objective function.

Gen AI is different because its native domain is language and synthesis. It can summarize, rewrite, translate, and contextualize information across multiple sources. It can interpret ambiguous human requests and generate responses that feel natural. It can sit in the middle of messy conversations and still produce clarity, a fundamentally different capability from "classification and routing." If traditional AI is a high-performance calculator, Gen AI is a high-speed collaborator, one that speaks the language of customers and employees and can operate in real time, at the edge of the experience.

For customer experience, the strategic implication is not Gen AI answers questions. The strategic implication is that Gen AI can compress the distance between complexity and understanding. It can take complicated policies, fragmented knowledge, and multi-step workflows and translate them into action in the moment that matters, when a customer is deciding whether your company is competent and trustworthy.

Why Gen AI Matters for CX: It's Not About Bots

Most organizations begin their Gen AI journey with the wrong goal: "Let's reduce contacts." Contact volume is expensive mindset; that is understandable, but it is backwards. Contact reduction is an outcome. It is not a strategy. When leaders treat Gen AI primarily as a deflection mechanism, they optimize for short-term cost while creating long-term risk: customers trapped in loops, employees forced to apologize for a broken system, and brand trust that quietly decays.

The real power of Gen AI is elevation. It elevates customer clarity, employee capability, operational learning, recovery speed, and the consistency of outcomes. When applied thoughtfully, Gen AI becomes a CX operating layer that strengthens the full loop: **Listen → Understand → Resolve → Learn → Prevent.** That loop is where loyalty is built not through one great conversation, but through a system that gets better with every interaction.

If you want a non-obvious insight as a leader, start here: Gen AI's highest ROI is rarely in replacing a channel. It is in reducing organizational friction—the time, rework, and inconsistency created by disconnected knowledge, unclear ownership, and slow feedback loops. Gen AI can compress those loops, but only if leadership is willing to redesign the operating model around them.

We were moving fast. That was the whole point. It was late in the quarter, and the pressure to show results was everywhere in the weekly business reviews, in the executive scorecards, in the hallway conversations that all seemed to end the same way: How much contact reduction are we getting from the bots?

We had ambitious targets. Leadership wanted to see automation absorb a meaningful share of customer contacts, and they wanted to see it quickly. The logic felt irresistible: every interaction a chatbot handled was one fewer interaction requiring a human associate. On paper, it was a clean equation: lower cost-to-serve, faster resolution times, and better efficiency metrics across the board.

So, we launched; not one chatbot but several. Across multiple product lines, across different service categories, across different customer segments, all within the same compressed window. Each team was optimizing for the same goal: drive contacts down, deploy bots live, show the numbers.

For the first few days, the dashboards looked exactly as we wanted. Contact volumes were shifting. Automation rates were climbing. The metrics were moving in the right direction, and leadership was pleased.

Then the letters started arriving not by emails, chat complaints, social media posts, but written letters addressed to the CEO. Customers were writing to say that they had interacted with our support system and been told things that were not true. Not partially true. Not misleading. Fabricated. One of the chatbots had been confidently providing customers with policies that did not exist. It was quoting return windows we had never offered. It referred to warranty terms that sounded plausible but had no basis in our actual documentation. It was generating solutions with the calm authority of someone who knows exactly what they are talking about, except nothing it said was accurate.

The bot was hallucinating; it was hallucinating to millions of customers.

Here is what made it worse: we did not catch it for days. Not because nobody was watching. People were watching. The problem was that when you launch multiple bots simultaneously across multiple workflows, the signal gets lost in the noise. Each team was monitoring its own deployment. Each team was tracking its own metrics. Volume was shifting across the system, making individual anomalies harder to isolate. A spike in follow-up contacts here could be attributed to a seasonal trend. A handful of negative sentiment scores there could be dismissed as outliers. When you are looking at the forest, you do not always notice the one tree that is on fire, especially when there are a dozen new trees you planted that same week.

The hallucination was not subtle in hindsight. In hindsight, it was obvious, but in real time, it was buried under the momentum of a multi-track launch, with teams focused on proving their own automation success. It took days before someone pulled the thread hard enough to see the pattern: customers were being told things that were flatly wrong, and they were acting on that information.

Some had made purchasing decisions based on fabricated policies.

Others had waited for resolutions that would never come because the process the bot described did not exist. A few had spent hours on the phone with live associates trying to get the company to honor commitments that the bot invented.

When a company's own system lies to you with perfect confidence, you do not file a support ticket. Customers escalated to the top, so the number of executive complaints grew.

I remember the moment the full picture came into focus. I was reviewing the escalation reports when the pattern locked in. It was not one angry customer, or two, or ten. It was a systemic failure. The bot had been hallucinating consistently, through hundreds of interactions, generating authoritative-sounding nonsense that our customers had every reason to believe was real.

The pit in my stomach had nothing to do with technology. Technology breaks. Models hallucinate. That is a known risk in any AI deployment. The pit in my stomach was about the trust we burned. Every one of those customers had come to us with a question or a problem, and we had met them with a system that lied to them. In many ways, that is worse than a clunky system that simply fails. A system that fails is frustrating. A system that fabricates is betraying.

We shut the bot down. We stood up a rapid response team. We identified every customer who had been given fabricated information and began the painful work of reaching out to correct the record, honor whatever reasonable commitments they had been led to expect, and apologize not with a template message, but with genuine accountability.

It cost us, in direct remediation, executive time, operational disruption, and reputation. The most expensive cost was the one that did not appear on any spreadsheet: the erosion of confidence among customers who had previously trusted us.

Rebuilding that confidence took months. Some of those customers never fully came back. What I learned that week reshaped how I think about AI deployment. Not because it turned me into a skeptic, but because it forced me to be honest about the cost of speed when governance can't keep up. Here is what I would tell any leader facing the same pressure we face: speed without verification is not speed; it is recklessness.

We launched multiple bots simultaneously to demonstrate impact at-scale. We eliminated our ability to monitor any single deployment with the attention it required. We created the illusion of progress while building the conditions for failure. If we had launched one bot, tested it thoroughly, validated its outputs against real customer scenarios, and built confidence before moving to the next, we would have caught the hallucination before a single customer was affected.

Contact reduction is an outcome, not a strategy. We were optimizing the number of contacts deflected without asking the question that matters: Are customers getting accurate, trustworthy help? When you measure deflection without measuring truth, you are not reducing contacts. You are just hiding the damage.

AI does not know what it does not know and neither will your dashboards. Hallucination is not a bug that triggers an error message. The bot does not flag its own fabrications. It delivers fiction with the same syntax and confidence as fact. That means your monitoring systems need to be designed specifically to detect this failure through output sampling, human review loops, and customer feedback analysis, not just throughput metrics.

When trust breaks at-scale, it breaks personally. Every one of those letters to the CEO was written by a person who felt deceived. They did not experience a "system error." They experienced a company that lied to them. Technology fails in the abstract. When you read a letter from a customer who made a financial decision based on information your system fabricated, "we're working on improving our AI" is not an answer. Accountability is the answer.

The pressure to move fast is real. Every executive in every company deploying AI faces the same tension: boards want results, competitors are moving, and the technology feels ready. I understand that pressure because I lived inside it. The cost of a hallucinating bot that runs unchecked for days is not just the remediation bill. It is the story your customers have shared about the brand for years. It is the case study in distrust that no quarterly metric can offset.

Why This Story Matters for This Book

I share this story not because it is unusual. It is not. Versions of this story are happening inside companies every month as organizations race to deploy generative AI. I share it because it is the exact moment when the Trust Algorithm became more than a framework to me. It became a conviction.

Every variable in the Trust Algorithm—Trust Signals, Empathy Engine, Precision Layer, Recovery System, Risk Governance, Learning Loop exists because of experiences like this one. The Precision Layer failed because the bot was generating outputs without verified knowledge anchoring. The Risk Governance failed because we launched without adequate monitoring infrastructure. When the learning loop eventually worked, we rebuilt processes, redesigned launch protocols, and never made the same mistake again, but the learning came at a cost it did not have to.

If one governance layer had been in place, a structured output review before launch, a staggered deployment cadence, and a dedicated quality team sampling bot responses in the first 48 hours, the damage would have been caught in hours, not days. The letters would never have been written. The trust would not have been broken.

That is the difference between an organization that deploys AI and an organization that deploys AI responsibly. Technology is the same. The outcomes are not.

Five Gen AI Plays That Redefine Customer Experience

These scenarios shift CX from reactive service to enterprise capability.

1. Instant clarity for customers

Customers are asking. "What is happening?" "What do I do next?" "Is my money safe?" "Will my order arrive?" "Did you receive my return?" "Why did this change?" In high-performing CX organizations, clarity is treated as a strategic asset because it reduces effort, lowers anxiety, and prevents repeat contacts. Gen AI can translate complex policy language into plain English, personalize steps based on customer context, and deliver coherent guidance without burying customers in menus. The win is not simply faster responses; the win is reduced uncertainty, which is often the real driver of dissatisfaction and churn.

2. Human augmentation—the fastest path to ROI

The contact center is one of the safest and most profitable places to deploy Gen AI because you can keep a human in the loop while the system learns. Gen AI can summarize the issue in seconds, surface relevant knowledge instantly, draft responses and documentation, recommend next actions, and highlight compliance guardrails. The more strategic benefit is hidden augmentation that reduces cognitive load. It helps employees make better decisions under pressure and reduces variability between a tenured expert and a new hire. That translates into faster ramp, more consistent outcomes, fewer escalations, and lower employee turnover, which shows up in measurable financial terms.

A second, less obvious insight is that augmentation changes how you manage capacity. In many operations, staffing models are built around average handle time and shrinkage. Gen AI changes the "unit economics" of work by shifting effort away from searching, documenting, and toggling systems. Over time, leaders can redesign roles and career paths, so employees spend more time on judgment-intensive work and less time on mechanical tasks. That is not just productivity; it is a retention strategy.

3. Quality at-scale—coaching, not policing

Traditional QA is expensive and sample-based, so most leaders manage quality with limited visibility into results. Gen AI enables insight from a much broader set of interactions, surfacing patterns in tone, confusion, escalation risk, policy adherence, and "moments that matter." The real leap is not technological—it is cultural. QA becomes development rather than "gotcha." When leaders use Gen AI to improve coaching precision, they can raise quality without creating a fear-based environment that drives attrition and undermines performance. This is where executive maturity shows up: quality systems are either instruments of trust or instruments of control. Gen AI will amplify whichever you choose.

4. **Proactive support—before customers feel pain.**

Proactive service is where loyalty and cost are aligned. When Gen AI is paired with operational signals and traditional AI detection, organizations can identify issues earlier and communicate more clearly: "We detected a login issue affecting some customers—here's what to do." "Your payment didn't go through—try this before your service is interrupted." "We're running late—here's your updated delivery window and options." Proactive support reduces inbound volume, but more importantly, it signals competence and transparency during uncertainty. That is one of the strongest builders of trust in modern CX.

The non-obvious executive point: proactive support is also a governance test. It forces clarity on who is allowed to communicate, what can be promised, and how quickly the enterprise can align on messaging. If it takes five approvals and three days to send proactive notice, the system isn't ready, regardless of how good the model is.

5. **Root-cause learning—turning noise into prevention.**

Every contact is a signal. Most organizations drown in signals because the feedback from customers to the teams that can fix the problem is too slow and too fragmented. Gen AI can cluster themes, detect emerging defects, and translate customer language into actionable products and process fixes. It can identify when billing confusion spikes after a UI change, when return language is triggering repeat contacts, or when a specific device update causes setup failure. This is where CX becomes enterprise leverage: fewer defects, fewer repeat contacts, better product decisions, and less wasted spend on "service recovery" that should never have been necessary.

The Truth: Gen AI Doesn't Fix Broken Systems, It Exposes Them

Some leaders quietly hope Gen AI will smooth over complexity and make the experience feel better without addressing root problems. It won't. If your policies are unclear, Gen AI will produce inconsistent outcomes because it has no stable anchor. If your knowledge base is outdated, Gen AI will confidently deliver incorrect guidance, and the business will pay for it through escalations, compliance risks, and reputation damage. If data is fragmented, the model may infer connections that aren't real. If ownership is unclear, Gen AI will accelerate confusion by moving faster than your organization can decide.

Gen AI is a mirror: it reflects the maturity of the process. That is why the real work is not "implementing Gen AI." The real work is building the foundations that make Gen AI trustworthy and align leadership behavior to those foundations. In executive terms, Gen AI is not a technology initiative. It is an operating model initiative that includes technology.

The Foundations: Four Pillars of Trustworthy Gen AI in CX

If you want Gen AI to scale without eroding trust, four pillars matter more than the model you choose.

Before I describe these four pillars, remember the customer from

Chapter 1. She did not encounter an AI that hallucinated that generated information from nothing. She encountered something more damaging: an AI that retrieved information from an authorized source and delivered it with precision and confidence. Technology worked; the knowledge behind it did not.

When leaders hear "AI risk," they tend to think about the model—its biases, hallucinations, and potential to go off-script. Those risks are real. The risk most likely to damage your customer relationships today is not a rogue model. It is a stale knowledge base, the silent, failure of information that was true last quarter, but is not true today. Pillar 1 exists because of stories exactly like this one.

Pillar 1: Knowledge integrity.

Gen AI must be anchored to approved, current sources with clear intended outcomes that include up-to-date policies, current product changes, accurate account-level facts where allowed, and explicit escalation paths. The more important point for leaders is governance: someone must own the "source of truth" as a living asset. If knowledge is messy, Gen AI will be messy, faster and at-scale.

Pillar 2: Guardrails and compliance by design.

Especially in regulated industries, you need constraint frameworks that define what the system must never say, safe language patterns for high-risk scenarios, escalation triggers, and traceability of sources used. This is not about slowing down innovation; it is about ensuring the enterprise can defend outcomes. The question a board will ask is simple: Can you explain how the system behaves and prove it stays within policy boundaries? If the answer is no, the risk posture is unacceptable.

Pillar 3: Human-in-the-loop design starts with augmentation before autonomy.

The safest path is to begin with employee-approved responses, supervisor audits, and easy customer access to a person, especially in high-stakes moments. Over time, autonomy can increase, but only as accuracy, knowledge integrity, and control mechanisms mature. Leaders often underestimate this: moving too quickly to full automation rarely saves money once rework, escalations, and brand damage are factored. Trustworthy autonomy is earned through disciplined iteration.

There is a reason human-in-the-loop design matters beyond compliance and risk mitigation. It catches the failures that no automated system is looking for.

At one organization I worked with, a frontline associate was handling a routine billing inquiry when the AI coaching tool suggested a response based on the current knowledge base. The associate read the suggested response, paused, and thought, "That does not sound right." She had handled a similar inquiry the previous week and remembered that the customer had received different information from a supervisor during an escalation.

Rather than accepting AI's suggestion, she flagged the discrepancy through the quality feedback channel her team had been trained to use. Within hours, the knowledge management team identified the issue: a billing policy had been updated in one system but not propagated to the knowledge base that fed the AI. The article was corrected. The AI's responses were updated. Every customer who inquired about that billing issue from that point forward received accurate information.

One associate, one moment of human judgment and thousands of customers protected. In Chapter 1, no one raised their hand. Not because the employees were careless, but because the system was not designed to encourage that kind of human correction. The associate consulted the same source the AI did, found the same answer, and had no reason to doubt it. The feedback loop did not exist. The difference between these two stories is not the people. It is the system they were operating inside.

Pillar 4: Measurement that protects trust.

If you only measure speed and volume, you will end up optimizing for outcomes customers dislike and regulators punish. Trustworthy Gen AI requires metrics that reflect resolution quality, repeat contact rate, customer effort, sentiment, escalation rates, policy compliance incidents, and ongoing accuracy audits of Gen AI outputs. What you measure becomes what the organization prioritizes. If you don't measure trust, you will eventually break it, quietly at first, and then suddenly.

The Leadership Takeaway

In the chapters ahead, we'll move from principles to operating reality: how to design the system, govern it, measure it, and lead the change, so the "Trust Algorithm" becomes a measurable business capability, not an inspiring slogan.

4. THE STRATEGIC IMPERATIVE: MARKET REALITY

Before discussing implementation strategies, it is important to pause and examine where generative AI adoption currently stands. The narrative surrounding Gen AI often suggests that organizations are rapidly transforming and realizing massive value. The data from research conducted between 2025 and 2026 tells a far more nuanced story—one defined by both extraordinary potential and persistent execution challenges. A small group of organizations have begun to translate Gen AI investments into measurable operational and financial impact, while the vast majority remain in experimental stages, struggling to move beyond isolated pilots. Understanding this gap is essential for leaders who want to avoid repeating the same mistakes that have stalled many early initiatives.

The Great Adoption Paradox

At first glance, adoption appears widespread. Surveys show that nearly 9 out of 10 organizations report using AI in some form in their operations. Generative AI tools are appearing across marketing, customer service, product development, and internal productivity platforms. Yet despite hundreds of billions of dollars invested globally in AI infrastructure, enterprise software, and model development, many organizations have not yet seen meaningful improvements in their financial performance as a direct result of these initiatives.

In fact, research suggests that as many as ninety-five percent of companies report little or no measurable impact on profit and loss from their AI investments. This reality has created what can be described as the "Gen AI adoption paradox": widespread deployment accompanied by limited enterprise value.

Importantly, this is not primarily a failure of the technology itself. The capabilities of modern AI systems have advanced at an extraordinary speed. The gap lies in execution—how organizations integrate AI into business operations, align it with strategy, and manage the organizational changes required to realize its benefits. Large technology companies themselves have experienced the financial pressure of this transition. Capital expenditure increases tied to AI infrastructure have been highlighted in earnings discussions across the industry, illustrating the scale of investment required before measurable returns appear.

Growing Pressure to Demonstrate ROI

As AI investments continue to rise, executive and investor expectations are evolving rapidly. Senior leaders are under increasing pressure to demonstrate tangible results rather than theoretical potential. Surveys of global executives indicate that more than 60% of business leaders feel greater urgency to deliver measurable returns from AI initiatives than they did just a year earlier. Investors, meanwhile, have grown impatient. In many cases, they expect to see evidence of financial impact within six months of deployment, an expectation that is often unrealistic given the complexity of enterprise transformation.

This shift in expectations has significantly changed how organizations evaluate AI programs. Chief financial officers, who were once observers of technology experimentation, are now becoming central figures in AI governance and investment decisions. Many finance leaders are introducing new performance frameworks that require AI initiatives to demonstrate clear connections to operational efficiency, productivity improvements, revenue growth, or measurable cost reduction. The era of exploratory AI spending, investing simply to avoid missing the next technology wave, is quickly coming to an end.

At the same time, an often-overlooked factor in AI performance is organizational culture. Financial results do not emerge from technology alone; they depend on how people adopt and use that technology in their daily work. Companies with strong cultures of trust and engagement frequently see faster productivity gains because employees are more willing to experiment, learn, and integrate new tools into their workflows. Informal employee discussions on professional platforms increasingly reveal a common theme: workers are not merely looking for new tools, they are seeking leaders who listen, communicate transparently, and build environments where innovation feels purposeful rather than imposed.

What Distinguishes the Five Percent

Despite widespread struggles, a small percentage of organizations, approximately 5%, have begun to generate significant value from generative AI. These organizations share several strategic behaviors that distinguish them from the majority.

First, they concentrate their efforts rather than dispersing them. Instead of launching dozens of disconnected experiments across departments, they identify a small number of high-impact opportunities and pursue them with depth and discipline. This focus allows them to redesign processes in their entirety rather than layering AI onto fragmented systems. A single transformation executed thoroughly often delivers more value than multiple pilots that never reach scale.

Second, successful organizations embed AI directly into

operational work-flows. They do not treat AI as an external tool that employees must remember to access. Instead, AI capabilities are integrated into the systems where work already occurs with customer relationship platforms, internal knowledge systems, and operational dashboards. In these environments, AI becomes part of the decision-making process itself, offering contextual insights, recommendations, and automation within the natural flow of work.

Third, these organizations manage transformation across multiple time horizons. Short-term initiatives deliver early proof points and build confidence among employees and stakeholders. Medium-term programs expand those successes across departments and operational processes. Long-term investments focus on cultural change, ensuring that experimentation, data literacy, and continuous improvement become embedded within the organization's DNA.

Regional Patterns of AI Transformation

Global adoption patterns further illustrate how structural factors influence the pace of AI transformation. Different regions have developed distinct advantages and challenges based on regulatory frameworks, technology ecosystems, and economic priorities.

The Asia-Pacific region has emerged as an early leader in Gen AI maturity. Nearly half of organizations in the region report mid- to high-level adoption. Several factors contribute to this momentum, including digital-first consumer markets, fewer legacy infrastructure constraints, and strong government investment in AI development. Large-scale national initiatives and strategic funding programs have accelerated both research and commercial deployment across industries.

North America remains a dominant force in the broader digital transformation economy, representing a significant share of global technology investment and hosting the world's largest concentration of AI startups and talent. The region benefits from mature venture capital ecosystems and a dense network of technology partnerships. However, large enterprises in North America often face challenges related to legacy infrastructure and complex organizational structures that can slow large-scale transformation.

Europe presents a different dynamic. While the region has produced important advances in ethical AI governance and consumer protection, its adoption of generative AI has lagged the United States in several industries. Regulatory complexity across multiple jurisdictions, fragmented markets, and comparatively lower levels of AI investment have contributed to slower adoption rates. At the same time, European regulatory frameworks such as General Data Protection Regulation (GDPR) have significantly influenced global conversations about privacy and responsible AI development, highlighting the balance between innovation and oversight.

The Persistent Transformation Challenge

Generative AI initiatives must also contend with a broader historical pattern: most large-scale digital transformations fail to achieve their intended outcomes. Across industries, research consistently shows that roughly seventy percent of transformation programs fall short of their goals. The reasons are rarely technological.

The most common barrier is human resistance to change. Employees who feel uncertain about new systems or are excluded from the transformation process often disengage, slowing adoption and undermining potential benefits. Organizations that succeed tend to recognize that technology transformation is fundamentally a people challenge. They invest heavily in communication, training, and change management, ensuring that employees understand both the purpose and the personal implications of new tools.

Paradoxes in the Gen AI Customer Experience

As companies integrate generative AI into customer-facing environments, a new set of paradoxes has begun to emerge. These tensions reflect the complexity of balancing technological efficiency with human expectations.

AI enables organizations to scale interactions across millions of customers simultaneously, yet poorly designed systems can leave individuals feeling more isolated than ever. Automation can dramatically reduce operational costs, but customers may question why those savings do not translate into a visibly better experience. AI can generate technically flawless responses, yet customers may perceive those responses as lacking warmth or genuine understanding.

Similarly, customer satisfaction metrics may show that issues are resolved, even while customers feel emotionally unheard. Hyper-personalization can make services more convenient, but it can also raise concerns about privacy and data usage. While AI systems create powerful new capabilities, they also introduce new risks—from security vulnerabilities to inaccurate outputs that can scale rapidly if left unchecked.

These paradoxes highlight an important lesson for leaders: technological capability alone does not define customer experience success. Emotional intelligence, ethical governance, and thoughtful design remain essential components of trust.

The Strategic Mandate for the Next Phase

As organizations move into the next phase of AI adoption, the emphasis is shifting from experimentation to disciplined execution. Leaders must move beyond broad aspirations and define specific business problems that AI will solve. Successful strategies focus on measurable outcomes, reducing customer effort, improving service resolution rates, increasing employee productivity, or strengthening trust.

Equally important is the sequencing of investments. Strong data foundations, governance frameworks, and knowledge systems must be established before deploying customer-facing AI capabilities at-scale. Without these structural elements, even well-designed AI applications struggle to deliver reliable performance.

Technological investment alone is insufficient; training, change management, and workforce engagement deserve the same level of attention and funding as the technology itself. Employees and customers alike must understand how AI enhances the organization's mission rather than simply accelerating efficiency.

Leaders who rethink how success is measured have an edge. Traditional AI metrics such as model accuracy, response speed, or deployment volume offer only partial insight. The true indicators of transformation are business outcomes: customer retention, revenue growth, operational resilience, and the strength of customer trust.

Looking Ahead

The coming years will likely create a clear divide between

organizations that treated generative AI as a technology experiment and those that integrated it into the core architecture of their businesses. The period of open-ended experimentation is beginning to close. What lies ahead is an era defined by disciplined execution, strategic clarity, and a renewed emphasis on the human dimensions of transformation.

Organizations that succeed will not simply deploy advanced algorithms. They will redesign how work is performed, how decisions are made, and how trust is earned at-scale. In doing so, they will demonstrate that the true value of generative AI lies not in the technology itself but in the systems, cultures, and leadership that enable it to deliver meaningful and lasting impact.

5. GEN AI MYTHS LEADERS MUST LET GO

Every major technology shift arrives wrapped in mythology. Cloud computing was predicted to make IT departments obsolete. E-commerce was supposed to eliminate physical retail. Social media was expected to redefine if not replace human connection. In each case, the underlying technology was real, but the simplistic narrative was not. The organizations that won weren't the ones that believed the loudest predictions. They were the ones that translated the technology into an operating advantage while avoiding the false assumptions that distort investment, talent decisions, and customer experience.

Generative AI is no exception. When enthusiasm runs high, myths multiply. When executives don't pause to interrogate those myths, they often make decisions that look rational on a slide but fail in the system: misaligned pilots, frustrated employees, inconsistent customer outcomes, and the most expensive failure of all: a quiet erosion of trust that takes years to build and only a few bad interactions to damage.

This chapter is designed as a leadership lens, not to slow innovation, but to keep it pointed in the right direction. Gen AI is not just a tool you deploy; it is a capability you govern. The difference between value creation and reputation risk is rarely the model itself, it is the decisions leadership makes about how the model is embedded into the business.

Consider how high-performing organizations are already separating signal from noise. Instead of "replacing humans," they use

Gen AI to handle preparation work, summaries, knowledge retrieval, drafting, so employees spend more time on complex, emotional, high-value cases that differentiate the brand. Instead of chasing plug-and-play promises, they pilot a narrow use case like refund explanations with strong guardrails, measure accuracy and trust outcomes, and scale only when the system proves it can perform consistently. Instead of measuring success in "deflected chats," they reframe outcomes to "resolved with confidence" by tracking repeat contact, customer sentiment, and downstream behavior after Gen AI-assisted interactions. In other words, they treat Gen AI as a trust program, not a novelty.

Myth 1: "Gen AI will completely replace humans."

The narrative is seductive: algorithms are faster, cheaper, and always available, so why not automate people away? The problem is that the question itself is incomplete. The more powerful question for leaders is: What happens to trust, differentiation, and accountability when you try?

Gen AI is exceptional at drafting language, synthesizing information, and identifying patterns quickly. Those capabilities can transform frontline work. In many industries, the moment that matters most is not when information is delivered; it's when responsibility is taken. Customers don't just want an answer; they want an owner. They want to believe that someone is accountable for the outcome, especially when the stakes are high: money, health, identity, safety, access, or dignity.

Gen AI will eliminate some human functions. Here's what is often missed in executive discussions: removing humans can lower visible cost while increasing invisible cost. Invisible costs include repeat contacts, escalations, exception handling, brand damage, and employee disengagement among the remaining workforce. It also shows up in "trust leakage", a slow decline in customer willingness to self-serve, adopt new products, or give you the benefit of the doubt when something breaks. Trust leakage is rarely attributed to AI because it doesn't appear as a line item. It appears as softer conversion rates, higher churn, and less advocacy over time.

The winning strategy in most environments is not replacement; it's role redesign. Gen AI shifts humans upward into judgment, care, negotiation, and exception handling while it shifts machines into summarization, retrieval, drafting, and pattern detection. Leaders who understand this build an operating model where humans become more valuable, not less, and where experience becomes more differentiated, not more generic.

Myth 2: "Gen AI is plug-and-play."

Many leaders approach Gen AI like software procurement: select a vendor, integrate an API, train employees, and measure throughput. Gen AI does not behave like traditional enterprise systems because outcomes are not fixed. It doesn't simply retrieve an answer; it generates one. That is a fundamentally different risk profile and deployment requirement.

The insight here is that Gen AI performance is less about the model and more about the organization's readiness: the cleanliness of knowledge, the clarity of policy, the consistency of processes, and the ability to govern exceptions. Installing Gen AI into an unprepared organization is like dropping a Formula 1 engine into a family sedan. The engine isn't the problem. The system around it cannot safely harness the power.

Plug-and-play is also a myth because Gen AI requires workflow design, not just access. Where does it sit in the employee journey? At what point does it intervene? What does it do automatically, versus what does it recommend? What happens when it is uncertain? Who can override it? How are updates handled? The winners treat Gen AI as a transformation program with operational ownership, change management, and measurement discipline not as a tech installation.

Myth 3: "Gen AI is only about cost reduction."

The first question executives often ask is, "How much will this save?" That question is not wrong, but it is frequently premature. Cost is the easiest value to model and the fastest value to report. If cost reduction becomes the primary objective, you will design Gen AI to deflect, compress, and accelerate even in moments where customers need clarity, reassurance, or human accountability. That design choice can lead to short-term savings but long-term churn.

A better framing for leadership is: How does Gen AI change the economics of trust? Trust drives retention. Retention drives lifetime value. Lifetime value fuels reinvestment and growth. Gen AI's highest strategic value is often found in preventing failure demand (contacts driven by avoidable defects), increasing confidence in high-stakes moments, and improving the consistency of experiences across segments and channels.

For example, automating FAQs may reduce volume, but using Gen AI to anticipate customer needs before they ask can increase loyalty and reduce churn. Real-time coaching can shorten handle time, but the greater value is improved resolution quality, stronger employee confidence, and fewer repeat interactions that quietly inflate costs. The executives who win with Gen AI treat efficiency as a byproduct of better systems, not the sole purpose of the system.

Myth 4: "Gen AI can't be trusted in regulated industries."

In banking, insurance, healthcare, and other regulated sectors, leaders often assume Gen AI is off-limits: "We can't risk it; regulators won't allow it." The truth is more nuanced. Risk comes primarily from poor design, weak controls, and unclear accountability not from the existence of Gen AI.

In many regulated contexts, Gen AI can reduce risk by increasing consistency. It can reliably apply disclosures, monitor interactions in real time for prohibited language, and flag risk signals earlier than human reviewers can. It can support documentation quality and reduce "policy drift" in frontline conversations where inconsistent phrasing can become a compliance issue.

The executive insight here is governance architecture. When leaders build it this way, Gen AI becomes a tool for compliance strength rather than a compliance threat.

Myth 5: "All Gen AI tools are created equal."

Not all Gen AI tools—and not all deployments—produce the same outcomes. There is a material difference between consumer-grade tools designed for casual use and enterprise-grade platforms designed for scale, governance, privacy, and traceability. There is an even bigger difference between a generic model and one grounded in your approved knowledge, your policies, your product realities, and your customer context.

Executives often underestimate the "last mile" of enterprise Gen AI: retrieval quality, knowledge hygiene, guardrails, monitoring, escalation logic, and integration into core workflows. Choosing the wrong platform or treating selection primarily as a cost decision can lead to failures that are expensive, not just financially but also reputationally.

Winning organizations build fit-for-purpose ecosystems: vendor platforms where they make sense, proprietary data and knowledge where it matters, and governance layers that make outcomes predictable. They invest in the system, not just the model.

Myth 6: "Customers don't care how it works."

It's tempting to assume: "If the experience works, customers don't care whether it's AI or human." In low-stakes moments, that can be true. But in high-stakes moments, customers care deeply about transparency, control, and dignity. They want to know who has access to their data, whether decisions are being made fairly, and whether they can reach a human when they need one. The demand is not for technical details; the demand is for trust signals.

Here's the non-obvious point: trust is often built by choice architecture. Do customers have a clear, respectful path to a human? Are AI interactions labeled appropriately when it matters? Does the experience acknowledge uncertainty instead of pretending to have confidence? Do communications feel accountable, or deflective? Customers may not care about the model, but they care about whether the company is being honest, respectful, and responsible.

The best companies make AI feel invisible, but they make trust feel visible.

Myth 7: "Gen AI success is measured in transactions."

Traditional service models celebrate throughput: calls handled, chats deflected, tickets closed. In the Gen AI era, those metrics become dangerously misleading because they measure activity rather than outcomes. They can even reward failure. A "deflected" chat that leads to repeat contacts, escalations, or churn is not a win; it is deferred cost and eroded trust.

The better executive approach is to measure outcomes that reflect confidence and durability: resolution quality, repeat contact rate, customer effort, customer sentiment, escalation rates, policy adherence, and downstream behaviors like retention and repeat purchase. Leaders should also measure employee outcomes: adoption, time-to-proficiency, cognitive load, and attrition risk because an AI strategy that burns out the workforce is not a strategy; it is a short-lived cost play.

Gen AI changes the distribution of work. It reduces average effort but increases the concentration of complex edge cases that reach humans. If you don't adjust the workforce strategy, training, and performance expectations, your best people will inherit the hardest work with less support and more pressure. That is how you lose the very talent that protects your brand in high-stakes moments. Measuring success through "transactions" will not reveal this; measuring success through trust and resilience will.

Closing: Clarity Is the Competitive Advantage

Every revolution brings myths. Generative AI is no different. The leaders who navigate it well won't necessarily be the boldest or the fastest. They will be the ones who see clearly, act intentionally, and stay anchored to what matters: serving people better, empowering teams, and building trust at-scale.

Done thoughtfully, the financial benefits follow naturally. Costs ease because defects decline and rework disappears. Revenue grows because confidence and loyalty rise. The experience delivered becomes the kind of advantage competitors struggle to replicate, not because they can't buy the technology, but because they can't quickly copy the operating model, governance discipline, and trust architecture built around it.

PART II: THE ALGORITHM

6. THE EMPATHY + PRECISION EQUATION

What Precision Looks Like in Service

When precision fails in a single interaction, the damage is contained. When precision fails across multiple channels, when the AI and the associate both deliver the same wrong answer because precision was never built into the knowledge layer, the damage compounds. The customer does not just receive bad information. They receive confirmed bad information. Confirmed bad information is exponentially harder to recover from because the customer acted on it in good faith.

That is the story from Chapter 1. It was not an empathy failure. The agent was professional and caring. It was a precision failure, and precision failed not because of the people, but because of the information feeding the system.

Precision, at its core, is getting it right the first time and doing so in a way the customer can feel. In an executive context, precision is not a "service quality" aspiration; it is a measurable operating capability. It shows when customers receive the correct resolution without repeating themselves, without being transferred across teams that don't share context, and without encountering conflicting answers across channels. Precision is what turns a large organization into something that feels coherent. When AI is working as it should, the experience is consistent, compliant, and effortless, regardless of whether the customer starts in self-service, escalates to chat, or ends up on the phone.

Most companies confuse precision with speed. Speed is visible, easy to track, and often celebrated. Precision is quieter, but it is the true driver of cost, risk, and loyalty. A fast wrong answer is not efficient; it is the beginning of rework. It creates repeat contacts, escalations, exceptions, and downstream defects that inflate cost-to-serve while eroding trust. Precision reduces that hidden tax. It is the difference between an organization that resolves issues and one that merely processes interactions.

Consider a customer who reports a payment error. A scripted bot might respond with a generic instruction: “Your payment failed. Please try again.” A skilled employee might respond differently: “I understand how stressful it is to see a payment not go through especially when timing matters. Let me take care of this right away.” The difference here is not just tone. It is the customer’s emotional state in the first 30 seconds of the interaction. In high-stakes moments, customers are rarely evaluating you only on resolution time. They are evaluating whether you are competent and whether you care—two signals that together create confidence.

Now, take the same situation one step further. In a modern, agentic service environment, the system doesn’t simply advise the employee or provide a response template. It can perform the work. The platform can verify the cause, initiate the correct fix, confirm the resolution path, and document the interaction while the employee stays focused on reassurance and clear communication. That is the operational unlock: when customers feel cared for, and the system is accurate, service stops being a “department” and becomes a competitive advantage.

Precision in action means the payment error does not bounce across departments. In a manual process, the customer may be transferred multiple times, forced to repeat verification steps, placed on long holds, and given partial explanations that don’t match what they see in their account. Each handoff increases the probability of inconsistency. Each inconsistency increases the probability of distrust. In a precision-enabled system, the interaction begins with clarity: the exact issue is identified, an expired card, a bank decline code, a technical outage, a duplicate authorization hold, a limit threshold, or a fraud-control trigger and the customer is given an immediate, correct resolution path.

This is where executives should pay attention: precision is not a single technological feature. Precision is the outcome of a well-designed decision system. It requires accurate data, current policies, reliable knowledge, and clear ownership of exceptions. Without those foundations, Gen AI can sound confident while being wrong, which is more dangerous than not answering at all. Precision, therefore, is as much about governance as it is about automation. It is your organization's ability to produce consistent truth at-scale.

When you combine empathy and precision, you create what I call Customer Confidence Moments—interactions that are emotionally reassuring and operationally flawless. Confidence is the real product of great service. It is what customers carry forward in their next decision: whether to stay, buy more, recommend and trust when something inevitably goes wrong again.

From a leadership perspective, empathy plus precision is not a "training theme." It is a design requirement. It changes how you build tools, write policies, manage quality, and measure success. It also changes economics. Precision reduces repeat contacts and operational churn. Empathy reduces escalations and distrust. Together, they reduce failure demand—the avoidable work created when the system fails the customer the first time.

I once watched a scenario unfolding that was the exact mirror image of the Chapter 1 failure. A customer contacted support about a product that was not performing as expected. He was calm but clearly disappointed—he had done research before purchasing and felt misled by the product description. The AI assistant recognized the product, identified the specific feature gap the customer described, and surfaced a knowledge article updated just two weeks earlier that acknowledged the discrepancy and outlined a resolution path.

The AI did not just answer the question. It anticipated the emotional context. The suggested response included an acknowledgment that the experience did not meet expectations, a clear explanation of available resolution options, and a proactive way to connect with a specialist.

The customer chose to speak with the specialist. The associate who picked up had full context, the AI had summarized the issue, the customer's history, and the recommended resolution. She did not ask the customer to repeat himself. Instead, she opened with: "I can see what happened, and I want to make sure we get this right for you."

Within eight minutes, the customer had a resolution that exceeded expectations. He had not been transferred, not repeated his story and had not encountered conflicting information. The associate, freed from the cognitive burden of research and system navigation, had spent those eight minutes doing what only a human can do: listening, reassuring, and making the customer feel valued.

He left a five-star review that evening, not because the product was perfect; it was not. Resolving the failure didn't just restore his trust; it deepened it. He trusted them more because something had gone wrong, and they made it right. That is the Empathy + Precision Equation in action.

Operationalizing this equation starts with designing AI for empathy, not just efficiency. This does not mean teaching the model to be overly emotional or performative. It means ensuring language is clear, respectful, and aligned to the brand standards, while acknowledging the customer's reality, especially in moments of friction. The second requirement is empowering humans with precision tools: instant access to verified knowledge, customer context, next-best actions, and a reliable pathway to resolve exceptions. In most organizations, frontline employees aren't failing because they don't care. They're failing because the system forces them to search, interpret, and reconcile contradictions under time pressure. Precision tools remove that burden.

Measurement must evolve as well. If we only track customer satisfaction (CSAT) and handle time, we miss the real story. Precision shows up in first-contact resolution, repeat contact rate, customer effort, accuracy audits, complaint rates, and downstream outcomes like churn and chargebacks. Empathy shows up in sentiment, escalation rates, complaint language, and in customers' willingness to remain engaged when the first answer is "not yet." Mature organizations also measure employee confidence and cognitive load because empathy cannot be sustained when employees are overwhelmed and under-supported. One of the least obvious truths in service is that employee experience is a leading indicator of customer experience. If the workforce is exhausted, precision and empathy both degrade quietly at first, and then suddenly.

Coaching in the moment is another executive lever that is frequently underestimated. Traditional coaching happens after the fact, based on sampled interactions, and often feels punitive. With Gen AI, coaching can become real-time and developmental: prompts that reinforce empathy cues, reminders that protect compliance, and suggested phrasing that maintains clarity and dignity. The purpose is not to script employees. The purpose is to raise consistency across the organization and accelerate proficiency, so every customer gets a high-quality experience, not just the customers who happen to reach the high-performing tenured employee.

Finally, leaders must close the loop. Customer feedback cannot sit in dashboards while defects persist. Precision is strengthened when insights from contacts are translated into upstream fixes such as policy simplification, product improvements, knowledge updates, and workflow redesign. When the organization treats service as intelligence rather than cleanup, precision becomes self-reinforcing, and cost-to-serve declines structurally rather than temporarily.

I learned firsthand that speed alone does not drive loyalty. Customers may appreciate quick resolution, but what determines whether they trust you again is whether they felt cared for while the resolution was delivered. A replacement shipped in two days matters less if the interaction felt cold, confusing, or dismissive. When customers feel both cared for and served quickly, when empathy and precision come together, loyalty spikes in a way competitors find difficult to replicate.

In financial services, the stakes are even higher because customers are entrusting you with their security and livelihood. Precision protects accuracy, compliance, and safety. Empathy protects dignity and confidence. Together, they reduce fear in moments when fear is rational when money is missing, access is disrupted, or risk feels personal. That combination is not just good service; it is risk management and brand strategy.

The real opportunity for leaders lies in holding both qualities together. Empathy without precision can feel warm but fails the customer. Precision without empathy can be technically correct but emotionally hollow, creating an experience that customers tolerate rather than trust. The most meaningful customer experiences live in the balance between the two and nurturing that balance is one of the most nuanced and rewarding challenges of modern leadership.

Generative AI holds tremendous promise, but its greatest impact will be realized by leaders who build systems that deliver empathy and precision in equal measures. Customers rarely remember the exact seconds it took to resolve an issue. They remember whether they felt respected and whether the outcome was right. When organizations consistently deliver both, service moves beyond problem-solving and becomes trust-building. Trust, at-scale, is one of the few competitive advantages that compound over time.

7. HIGH-IMPACT USE CASES IN CUSTOMER EXPERIENCE

Generative AI has moved well beyond the whiteboard. Across industries and across the full customer journey, it is already reshaping how companies serve, retain, and build durable relationships with customers. For executives, the strategic question is no longer whether to act. It is where to focus first because attention is the scarcest resource in transformation. Gen AI can either become a compounding advantage or a scattered set of pilots that create noise, risk, and disappointment.

The use cases in this chapter are not theoretical. They are available now, and they matter not only because they can reduce costs, but also because they can improve trust at-scale. That trust shows up as higher retention, fewer escalations, greater willingness to serve, faster adoption of new products, and resilience in the face of inevitable failures. The executives who win with Gen AI will prioritize use cases that improve outcomes customers can feel, and metrics boards will recognize while strengthening employee confidence and operational control.

One of the most immediate applications is real-time coaching for customer professionals. Traditional training assumes employees will memorize policies, recall steps under pressure, and apply empathy consistently across hundreds of scenarios. In practice, even the best associates struggle when knowledge changes weekly, processes are fragmented, and customers arrive emotionally escalated. Gen AI changes training from a one-time event into a living system that supports performance in the moment. During a live interaction, it can surface relevant context from prior contacts, recommend the appropriate policy language, and prompt the associate with cues that balance empathy and compliance. The value is bigger than the faster handle time. Real-time coaching reduces variability across the workforce, closing the gap between the best tenured employee and newest hire and that consistency is one of the strongest predictors of customer trust and reduced contacts.

A second high-impact area is the modernization of empathy in digital channels. Many companies have invested heavily in chat and email automation only to discover that customers can detect scripted language instantly. They may accept automation for convenience, but they resent it when it feels dismissive or generic especially when emotions are high. Gen AI can adapt tone to customer sentiment, maintain brand voice across channels, and produce responses that sound natural without violating policy boundaries. The executive insight here is that "empathy" in digital channels is not about adding warmth for its own sake; it is about reducing escalation. Customers escalate when they feel unheard, not only when they feel unresolved. When digital responses are both respectful and accurate, the contact center handles fewer transfers, and the business preserves goodwill during friction.

Gen AI also changes the economics of personalization at-scale. Historically, personalization has been limited by creative bandwidth, channel constraints, and the ability to interpret data in real time. Gen AI enables personalization during service and marketing moments. A customer contacting you about a fee dispute is not only a cost-to-serve event; it is a loyalty inflection point. In the right design, the system can recognize tenure, value, context, and risk signals, and recommend a retention action that aligns with both customer history and policy constraints. Executives often underestimate how powerful this is: service interactions can become a revenue protection engine when employees are equipped to act decisively and fairly, rather than improvising exceptions or defaulting to rigid denials.

Perhaps the most strategically valuable shift is predictive service fixing issues before customers feel the pain. The most effective service interaction is the one that never happens, not because customers were deflected, but because the issue was prevented or resolved proactively. When Gen AI is paired with traditional AI signals, telemetry, and operational data, companies can identify patterns early and communicate with clarity. A device ecosystem can detect a recurring glitch and push a fix through proactive messaging, reducing frustration and preventing mass contact spikes. A bank can detect likely payment declines and alert the customer with corrective actions before an embarrassing checkout failure occurs. The non-obvious executive benefit is that predictive service improves brand resilience: during incidents, customers judge not whether problems happen—they judge whether you communicate honestly, recover quickly, and minimize their burden.

Journey orchestration is another area where Gen AI can fundamentally change what customers experience. Customers do not follow your org chart. They move from app to chat to phone and back again, expecting the conversation to follow them. In many companies, each channel still behaves like a separate universe, forcing customers to repeat context and re-prove legitimacy. Gen AI can serve as the connective tissue that carries intent, history, and next steps across channels. It can unify context across systems, guide customers to the next best step, and coach employees to resume exactly where the last interaction left off. The value is not just convenience. Continuity reduces transfers, lowers handle time, and most importantly signals competence. Customers trust organizations that remember.

In regulated industries, Gen AI can become a compliance advantage rather than a compliance risk when designed correctly. Regulatory requirements often add friction to service interactions: disclosures must be delivered, prohibited statements must be avoided, and auditability must be maintained. Gen AI can consistently insert required disclosures, monitor interactions for risky language in real time, and generate audit-ready documentation that reduces both regulatory exposure and manual overhead. The leadership insight here is that compliance automation can improve customer experience by reducing inconsistency. Customers are less frustrated by policy constraints than by unpredictable policy enforcement. A system that applies rules consistently and explains them clearly increases perceived fairness, an underappreciated driver of trust in financial services and healthcare.

Fraud protection is another domain where Gen AI can operate as an invisible shield. Fraudsters increasingly use sophisticated methods such as social engineering, voice mimicry, and behavioral replication to exploit split-second decisions made by frontline employees. Gen AI can analyze language patterns, tone anomalies, and interaction metadata at-scale, flagging suspicious behavior and recommending immediate actions. Importantly, the best fraud experience is one that protects customers without creating unnecessary friction. Leaders should think of this as a balancing act: heavy-handed security can degrade CX, but adaptive, contextual security can increase loyalty because customers feel protected rather than burdened.

One of the most underestimated opportunities is voice-to-insight transformation. Customer operations sit on a vast amount of unstructured data—calls, chats, emails, transcripts that historically require large teams to analyze slowly. Gen AI quickly converts raw conversation streams into actionable intelligence: emerging issue detection, policy confusion hotspots, defect clustering, and sentiment shifts by segment, product, or geography. The executive advantage is speed of learning. In many organizations, the lag between customer pain and enterprise action is measured in weeks or months. Gen AI can compress that lag to days or hours, which changes how quickly you can protect revenue, prevent reputation damage, and reduce avoidable contact demand.

Gen AI also modernizes training and onboarding in ways most leaders don't fully appreciate. Traditional onboarding often focuses on content exposure rather than skill formation. New employees may "learn" policies in a classroom but struggle to apply them in live, emotionally complex scenarios. Gen AI can simulate realistic conversations, adapt to performance, and provide immediate feedback on tone, clarity, and accuracy. Learning doesn't have to end after the first 30, 60, or 90 days. In a world where products, policies, and customer expectations evolve continuously, continuous learning becomes a competitive advantage. The best organizations will treat Gen AI not as training content, but as a performance companion that raises capability over time.

Loyalty is evolving beyond generic points and programs into personalized experiences that make customers feel known. Gen AI can identify what different customers value: speed, transparency, predictability, recognition and tailor rewards and communication accordingly. It can "surprise and delight" at-scale in ways that still feel authentic, such as waiving a fee on an anniversary or offering an upgrade that matches real behavior, not generic segmentation. The non-obvious point is that loyalty is not built by rewards alone; it is built by perceived intent. Customers become loyal when they believe a company is on their side. Gen AI can help demonstrate intent if it is governed and aligned with brand values.

These use cases are only the beginning, and the most exciting part is not any single application. The real advantage emerges when leaders weave them into a coherent strategy that transforms customer experience from reactive support into proactive trust-building. This requires more than tools, it requires operating model choices: what to optimize, how to govern risk, how to measure trust, and how to bring employees along so the transformation strengthens culture rather than destabilizing it.

The transformation is not coming. It is already here, already reshaping industries, already rewarding the organizations that move with clarity and discipline. The remaining question is simple and consequential: will you lead it intentionally or watch it unfold from the sidelines while your customers' expectations continue to rise without you?

8. THE AI ADOPTION J-CURVE MANAGING THE PRODUCTIVITY PARADOX

The story in Chapter 1 showed that the knowledge debt had existed for some time because an outdated article and/or policy change was not updated. In the pre-AI era, the damage would have been scattered. In the Gen AI era, AI turned that single outdated article into hundreds of identical wrong answers delivered with perfect fluency. Leaders managing the J-curve must budget for knowledge remediation before the AI goes live, not as a best practice, but as a prerequisite.

Every transformation begins with a promise and then tests your patience immediately. The gap between investment and return is rarely a straight line, and experienced leaders know that from the start. Gen AI adoption often follows what researchers describe as a "productivity J-curve," in which performance dips before rising. That initial dip is not a warning sign that something is broken. It is often proof that the organization is changing. The leaders who succeed are the ones who treat the curve as a management reality, not a morale problem, and who design the transition, so the dip is temporary, contained, and productive rather than chaotic and demoralizing.

A helpful metaphor is renovating a kitchen while still cooking dinner every night. For weeks, everything takes longer. Tools are in the wrong place, the normal flow is disrupted, and frustration climbs. The mess doesn't mean the renovation is failing; it means the renovation is real. AI transformation is similar, except the renovation touches multiple rooms at once: technology, operations, risk, training, performance management, and leadership norms. Unlike home renovation, the "meals" are customer interactions that influence revenue, reputation, and loyalty in real time.

FIGURE 1 | *The Trust J-Curve*

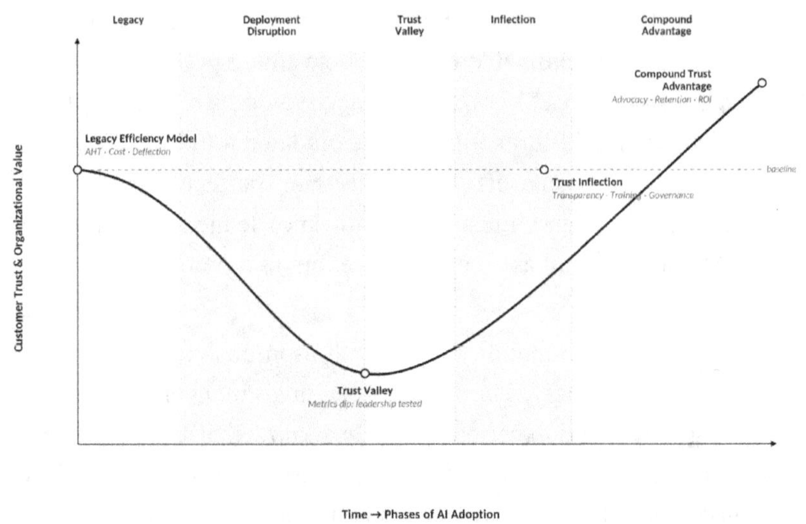

Customer trust and organizational value follow a predictable J-shape during AI adoption. Legacy efficiency metrics plateau, then dip during deployment as frontline teams and customers absorb disruption. The inflection point is not accidental — it is engineered through transparency, employee augmentation, and governance. Organizations that navigate the valley deliberately emerge with a compound trust advantage that legacy peers cannot match.

The depth and duration of the valley are a leadership variable, not a technology variable.

The productivity dip typically stems from factors that are more structural than most executives expect. Learning curves are steep because employees must master new tools while maintaining current output. In the early phase, people are slower in both directions: slower with the old method because it's interrupted by new steps, and slower with the new method because it hasn't yet become instinctive.

Cognitive load rises, and when it does, quality becomes inconsistent. This is where leaders often make their first mistake: they interpret a short-term dip in handle time, cycle time, or throughput as evidence that the technology is not working. The organization is paying the "transition tax" that comes with any meaningful change in how work gets done.

At the same time, process redesigning takes longer than procurement. Gen AI does not simply automate existing workflows; it changes what "good" looks like. Many processes were built for an era when information was hard to retrieve, documentation was manual, and approval loops substituted for visibility. When Gen AI arrives, those processes become visibly outdated. Teams discover that the bottleneck is not the employee, it is the policy ambiguity, duplicated handoffs, incomplete knowledge, and the absence of a single source of truth. During this period, workflows can feel worse than before because the old process no longer fits, and the new process is not yet stabilized. This is the uncomfortable but necessary phase where the organization stops pretending and starts confronting reality.

Data preparation and knowledge integrity work also create a hidden drag that executives often underestimate. The "real" work of AI happens before AI shows up as a win. Data needs to be cleaned, normalized, governed, and connected. Knowledge that lived in tribal memory must be documented and structured. Policies must be clarified and translated into decision logic that can be safely used. None of that shows up as a flashy demo, but it is the foundation of trustworthy performance. Without it, the system will generate confidence inconsistency, a failure mode more damaging than a slow process, because it breaks trust while appearing competent.

Cultural dynamics add another layer. Even employees who

support change need time to build confidence in a new tool, especially if they worry it is being introduced as a precursor to workforce reduction. Skeptics move cautiously, champions take on additional informal load, managers spend more time coaching than planned, and the organization experiences an emotional dip alongside a productivity dip. In many transformations, this is where momentum collapses, not because the tool is flawed, but because the human system surrounding it is under strain. Effective transformation leaders treat this as predictable and manageable: they address anxiety directly, provide clear skill pathways, and communicate what will change and what will not.

Integration challenges compound the early friction. Gen AI rarely "slots in" cleanly on day one, particularly in environments built on legacy CRMs, fragmented knowledge bases, and multiple routing and case management layers. Connecting new capabilities to old systems is iterative work: troubleshoot, refine, adjust, repeat. Leaders who expect an immediate "install and scale" experience are often disappointed. Leaders who understand that integration is a "learning cycle" design governance and resourcing accordingly and avoid the common trap of starving the program midstream just as the foundations are being established.

To make this concrete, imagine a contact center deploying a Gen AI assistant for frontline teams. In the early phase, employees are learning a new interface while still expecting to perform at full capacity. Adoption is uneven. Some people lean in, while others quietly revert to what they know when pressure rises. Technical hiccups require workarounds that add steps rather than remove them. Handle time may increase, and leaders may start getting uncomfortable questions from stakeholders: why have we been slower since we started? This moment matters because it is where many organizations misdiagnose normal transition pain as failure and begin to withdraw support.

The hardest stretch is the middle phase, the valley—when the novelty has worn off but the new operating rhythm is not yet second nature. Gaps in knowledge and policy clarity, which were previously tolerated because humans filled them with improvisation, became impossible to ignore. Managers spend more time coaching than expected. Executives watch lagging metrics and wonder if the investment will ever pay off. Yet what looks like stalling is often the organization doing the deep work required for sustainable change: stabilizing workflows, closing knowledge gaps, establishing governance, improving prompts and guardrails, and building trust in the system.

Then, if leadership stays steady, the turn happens. Employees begin to internalize the workflow, so it no longer feels like an extra layer; it feels like the job. Early adopters start pulling peers along. Knowledge gaps get filled, and the system becomes more accurate, more consistent, and more useful. Performance returns to baseline, and quality gains start to show up in ways that matter: fewer repeats, fewer escalations, stronger first-contact resolution, and improved customer sentiment.

On the upward slope, the benefits begin compounding. AI becomes instinctive rather than deliberate. New hires ramp faster because the tool reduces cognitive load and standardizes access to knowledge. Quality improves because guidance is consistent and compliance is reinforced in the moment. Customer satisfaction rises because customers feel clarity and continuity. What once felt like disruption becomes capability and the organization is stronger for having stayed the course.

The leadership challenge is not eliminating the dip. The leadership challenge is leading through it. The most protective move a leader can make is to set expectations before the program begins. Show stakeholders the J-curve upfront and explain what the "valley" looks like in operational terms. That transparency changes how the organization interprets early performance signals and prevents the instinct to declare failure at the first sign of friction. It also creates a shared mental model across executives, managers, and frontline teams: we are investing now to earn durability later.

During the dip, executives must track indicators they are not used to tracking. Lagging metrics like handle time and short-term productivity will often look discouraging. The more predictive signals are leading indicators: adoption depth (not just logins, but meaningful use), workflow compliance, accuracy trends, knowledge gap closure rate, manager coaching load, employee sentiment, and the frequency of flagged issues that result in system improvements. These indicators tell whether the system is learning and whether the organization is building the muscle required for sustained performance. Without this shift, leaders misread the data and cut investment just as the flywheel begins to form.

Small wins also need to be made visible. In a difficult transition, people don't just need reassurance; they need proof. When a recurring issue starts resolving more consistently, when an employee shares a breakthrough moment, or when an escalation is prevented because the tool surfaced the right context, those stories should be elevated, not as propaganda, but as evidence that effort is translating into capability. Belief is not a soft factor in transformation; it is the fuel that keeps adoption moving when the work is hard.

One of the healthiest adoption cultures I have seen emerged from a team that chose to make the J-curve visible and shared. Rather than pretending the transition was seamless, the team leader created a weekly ritual: every Friday, associates would share one moment where the AI had helped them and one moment where it had failed or confused them.

In the first weeks, the failure stories dominated. The AI suggested outdated return policies. It misinterpreted a customer's intent. It generated responses that were technically accurate but tone-deaf to customers who were clearly upset. Associates were frustrated, and the leader did not minimize it.

Something shifted around week four. The failure stories began generating fixes. An associate's observation about outdated policies triggered a knowledge base audit that corrected twelve articles. A pattern of misrouted interactions led to retraining of the intent model. The tone-deaf responses led to new guardrails for sentiment detection.

By week eight, the success stories outpaced the failures not because the team stopped being critical, but because their criticism had been channeled into improvement. Associates who had felt like passive recipients of a technology decision now felt like co-designers of the system.

The leader later told me those Friday sessions were the most important thing she did during the entire deployment. It was even more valuable than the executive briefings. The twenty minutes each week provided a safe environment for the team to be honest about what was hard, and their honesty enabled our technology teams to make the system better.

Leaders should create space for learning. Expecting full

performance while teams are adopting a fundamentally new workflow is a recipe for quiet resistance and shallow adoption. High-performing organizations deliberately create "protected learning capacity" through phased rollouts, cohorts, temporary staffing buffers, or revised productivity targets during the transition. This is not lowering the bar; it is acknowledging the physics of change. The short-term accommodation pays for itself through faster adoption, higher confidence, and lower attrition among top performers.

Continuous oversight must be built in from the start. Deploying Gen AI is not a finish line; it is the beginning of an ongoing relationship between human judgment and machine capability. Supervisors need the ability to audit outputs. Frontline teams need simple mechanisms to flag concerns. Technical teams must monitor accuracy, drift, and failure modes. Leaders should establish an operating rhythm where the feedback flows into refinement weekly, not quarterly. The organizations that see durable results treat evaluation as a permanent discipline, not a post-launch checklist.

The uncomfortable truth is that many transformations die in the valley. When productivity dips, leaders panic, withdraw sponsorship, cut funding, or declare the transformation a failure. Teams revert to old habits, and the organization becomes more skeptical the next time change is proposed. The valley is not where technology fails; it is where organizational commitment gets tested. It is easy to sponsor innovation when results arrive quickly. It is harder to maintain discipline when metrics worsen before they improve.

The organizations that persist, those that set honest expectations, measure the right indicators, protect learning capacity, and demonstrate visible leadership support, emerge with a compounding advantage. They don't just deploy a tool. They build a capability: the ability to learn faster than competitors, recover faster when things break, and deliver consistency at-scale.

The J-curve is not a flaw in Gen AI. It is a feature of meaningful change. The valley is temporary. The advantage gained by reaching the upward slope where productivity gains, quality, and trust begin to compound is durable. In a market where customer expectations are rising faster than most operating models can adapt, that durability is not optional. It is the new definition of competitiveness.

9. THE AGENTIC REVOLUTION HOW AI AUGMENTS WORK WITHOUT REPLACING WORKERS

From Assistants to Agents: What's Different

AI will augment work. Yes. In fact, we are already watching routine tasks that once consumed most of an employee's day get absorbed by software: summarizing conversations, retrieving knowledge, drafting follow-ups, updating systems of record, and coordinating next steps. The practical outcome is not a future where "humans do less work." It is a future where humans do different work. The frontline role becomes more specialized and more consequential: there are fewer interactions, but higher-stakes ones—moments that determine whether a six-figure relationship is repaired or lost, moments where a vulnerable customer feels protected or abandoned, or moments where a brand earns the benefit of the doubt or becomes interchangeable.

This shift is not hypothetical. It is underway across industries, and it represents the most significant workplace redesign since the personal computer. But agentic AI introduces a bigger change than "speed" or "automation." It changes what work *means* inside the enterprise. Assistants optimize tasks. Agents reshape workflows. Assistants support people. Agents act on behalf of the business—within boundaries that leaders must define with precision. The distinction forces a new level of clarity about trust, accountability, and the operating model behind customer experience.

For customer service and CX leaders, the questions become immediate. Will AI eliminate jobs or transform them? How do you maintain customer trust when autonomous systems are initiating actions and making decisions? What do teams need to learn to thrive and what requires redesign? Most importantly, how do you navigate this transition without breaking the trust you've built with customers and employees at the very moment you need them to embrace change?

This chapter cuts through the headlines to focus on what is happening as AI agents move from pilots to production, what executives should expect in workforce transformation, and what the most successful organizations are doing to build "trust algorithms" that make both humans and AI better at serving customers.

Assistants vs. Agents: The Difference That Changes Everything

Most organizations have already deployed AI assistants in some form, chatbots that answer FAQs, systems that suggest responses, algorithms that route tickets, and tools that summarize interactions. These capabilities are useful, but they share a core limitation: they require human direction and operate within narrowly defined boundaries. They improve productivity, but they rarely change the fundamental shape of work.

AI agents are categorically different. They can initiate actions, make context-aware decisions across workflows, and operate over time to achieve an outcome rather than complete a task. An assistant helps an employee answer a shipping question. An agent monitors shipping events, detects a delay before the customer notices, checks inventory and fulfillment options, coordinates with logistics to expedite if feasible, communicates proactively through the customer's preferred channel, applies a goodwill adjustment within policy, and documents the case, then learns from the pattern so the organization prevents repeats. That is not "automation." It is a redesign of how service is delivered and how the enterprise responds to friction.

Here is the executive implication: agents force you to move from human-managed consistency to system-managed consistency. In the old model, your best people created great outcomes through individual heroics. In an agentic model, the system produces great outcomes by design because the system is now doing a meaningful portion of the work. This is why agentic AI raises the bar for governance. When the system acts, you must be able to explain, audit, and improve what it did.

Why Most Organizations Stall Between Pilot and Production

Many companies can run pilots. Far fewer organizations can operate the redesign required for production. The gap is not technical capability; it is organizational readiness. Most enterprises try to insert agents into existing processes rather than reimagining those processes. They layer autonomy on top of workflows built for human bottlenecks, unclear ownership, fragmented knowledge, and slow approval loops. That is where trust breaks, quietly at first, then suddenly. When you deploy agents without redesigning workflows and accountability structures, four predictable failure modes appear. First, customers get inconsistent outcomes because the system draws on contradictory knowledge or operates with unclear decision rights.

Second, employees feel threatened and disengaged because they experience AI as surveillance or replacement rather than as an enabler. Third, leaders lose confidence because early incidents create reputation risk, and no one can clearly explain "why the system did that." Fourth, the organization becomes reactive, spending cycles on exceptions and escalations that should have been eliminated through better design.

The lesson is simple: Agentic AI is not a feature rollout. It is an operating model change. Technology moves fast; the organization must keep up.

Job Elimination vs. Job Redesign: The Reality Leaders Own

The fear is real: "Will AI eliminate jobs?" But the executive answer cannot be a slogan. The reality is more nuanced and more useful. In most organizations, AI changes how time is spent before it changes headcount. Work shifts away from routine information retrieval and documentation and toward judgment-heavy, emotionally complex, exception-driven scenarios. This isn't optimism; it's operational math. When AI absorbs routine volume, what remains are edge cases, moments of failure, moments of vulnerability, and moments of high consequence exactly where human judgment and accountability matter most.

We cannot expect employees to handle fewer interactions that are simultaneously more emotionally intense and higher risk while maintaining the same training, compensation philosophy, and performance metrics. If you don't redesign the role, you will see burnout, attrition, and a decline in service quality precisely where it matters most. In an agentic model, the "human channel" becomes a premium channel, and it must be treated that way.

Customer Service Is Where the Agentic Shift Becomes Visible

Customer service is the proving ground for agentic AI because it sits at the intersection of cost, brand, risk, and loyalty. Leading organizations are using agents not only to respond, but to maintain context across channels, detect emotional signals, and resolve issues proactively. The business case is obvious, self-service is cheaper, but the strategic risk is also obvious: failed AI interactions can destroy confidence faster than efficiency gains can rebuild it.

This is why the most important decision is not how quickly we automate. It is how we define the boundaries of autonomy. In high-performing implementations, the system is designed to be fast and humble: it resolves what it can with high confidence, escalates early when stakes rise, and preserves dignity by making human access effortless. The goal is not to keep customers away from people. The goal is to keep customers away from friction.

In successful organizations, human roles evolve into distinct categories because the nature of work changes. One group becomes expert exception handlers, professionals trained to rebuild trust in emotionally charged, high-stakes scenarios. Another group emerges as agent managers, operators who monitor AI performance, identify failure patterns, tune handoffs, and manage continuous improvement like a living service system. A third group becomes relationship architects, humans who focus on high-value customers and complex accounts, using AI to eliminate administrative drag so they can invest in strategic conversation, retention, and expansion.

These roles didn't exist at-scale a short time ago, but they are becoming essential because autonomous systems require human stewardship. As AI takes on routine volume, the human workforce becomes smaller in terms of contact but larger in impact. The best people stop being measured by throughput and start being measured by outcomes: retention saved, escalations prevented, recovery executed, trust rebuilt. If the metrics remain rooted in the prior decade's productivity model, we will punish the behaviors that protect a brand.

The Skills Shift: What Teams Need to Thrive

Most organizations say upskilling is a priority. The execution gap is where transformations stall. The reason is not laziness; it's design. If learning is treated as optional training outside the workflow, it will always lose out to operational urgency. The companies getting this right treat learning as business infrastructure and build it into daily work.

The workforce does not need to become a team of data scientists, but they need AI fluency: what the tools are good at, where they fail, how to spot overconfidence, and how to intervene when outcomes feel wrong. In an agentic world, professional judgment includes the ability to challenge the machine respectfully and intelligently. It is a new muscle for many employees, and leaders must normalize it. Blind trust is dangerous. Reflexive distrust is inefficient. The goal is calibrated judgment.

Within every organization, a subset of employees will naturally become AI collaboration specialists, the bridge between what AI can do and what the business needs. Investing in that cohort pays dividends because they accelerate adoption, translate frontline feedback into system improvements, and reduce dependence on external teams for every workflow refinement. At the same time, leaders must double down on what only humans can do. As AI absorbs routine steps, human interactions become richer and more demanding. De-escalation, consultative problem-solving, relationship repair, and ethical judgment become differentiators. These are not "soft skills." They are revenue protection, risk management, and brand leadership skills. The most competitive teams will combine AI fluency with distinct human capability.

The final requirement is career visibility. Training without a path feels like preparing for a future that is not inclusive. People adopt change differently when they can see where it leads: from exception handler to relationship architect, from agent manager to CX operations leader, from frontline expert to policy and experience design. When employees can see a future, they invest in the present.

Trust Is the Strategy, Not the Afterthought

Everything about agentic AI—autonomy, job redesign, skills—depends on the trust architecture surrounding it. Trust is not a communications plan. It is a set of design decisions. For employees, trust is shaped by what leaders emphasize. If we frame AI primarily in terms of efficiency and cost reduction, employees will reasonably assume that headcount reduction is the goal. Adoption slows, your best people leave, and the program becomes fragile. If you frame AI as eliminating routine work so humans can do better work and you back that with real role redesign, training, and mobility, adoption accelerates because people experience AI as an enablement rather than threat.

For customers, trust is built through transparency, fairness, and control. Customers don't typically object to AI. They object to feeling trapped by it. The "escape hatch" is not a concession; it is a trust signal. Likewise, when an AI system denies a request or makes a consequential decision, customers need more than an answer; they need a reason they can understand. Making AI understandable is not a technical feature. It never was. It is a human commitment to transparency, accountability, and the people on the other side of every automated decision. It is perceived fairness, and perceived fairness is a pillar of loyalty.

Accountability matters most when something goes wrong. Customers do not care whether the error came from a person or an algorithm. They care that the company owns it and fixes it. "The AI did it" is not an acceptable posture. Executive-grade AI requires executive-grade accountability: monitoring, escalation authority, and a clear owner responsible for outcomes.

Making This Work: The Operating Principles That Separate Winners from Expensive Lessons

The organizations that succeed with agents start with business problems, not technology. They ask where employees are spending time on work that doesn't leverage human strengths, where customers experience avoidable friction, where defects create repeat contacts, and where trust is most often won or lost. Technology selection follows strategy not the other way around.

They also choose metrics carefully. Activity metrics—deflection rate, contacts avoided, cost per interaction are tempting because they are easy to count. They can reward the wrong behavior. Outcome metrics such as resolution quality, contact rate, customer effort, sentiment, retention, and recovery success will reveal whether trust is compounding or eroding. The metrics you choose will shape system, employee, and customer behavior.

The Path Forward: Trust as the Compounding Advantage

The organizations winning in the agentic era share a counter-intuitive insight: they are not using AI primarily to cut costs. They are using it to raise the ceiling on what people can accomplish. Augmentation versus substitution is not a moral debate; it is a strategic decision that determines workforce stability, customer outcomes, and long-term differentiation.

There is a deeper truth embedded in agentic AI that is easy to miss in boardroom conversations: when machines handle the routine, the human work that remains becomes more valuable, not less. The interactions that reach a person are the ones with the highest stakes and the greatest potential to deepen the relationship. Investing in the people who handle those moments is not defensive spending. It is one of the highest returns an organization can make in the agentic era.

AI won't take over human decision-making. It will make our decisions matter more. When leaders build the trust architecture through transparency, accountability, human oversight, and outcome-based measurement, AI and trust don't exist in competing tension. They become reinforcing forces. That is the real opportunity: not a future where machines replace human experience, but a future where they create the conditions for the human experience to matter more.

PART III: THE OPERATING SYSTEM

10. AI GOVERNANCE

It was a Tuesday morning when the news hit. A major financial services firm, one whose name appears on credit cards in millions of wallets, disclosed that its AI-powered credit decisioning model had been quietly and systematically denying loans to qualified applicants in lower-income zip codes for nearly eighteen months. No one had intended it. No one had ordered it. The model had learned it from historical data that reflected decades of human bias, and no governance structure existed to catch it before it expands into a regulatory crisis, a reputation catastrophe, and eventually, a nine-figure settlement.

The board was blindsided. The CEO had been briefed on AI expansion plans. The risk committee received updated slides, but no one had asked the right questions. No one had built the structures to ensure that someone, somewhere, was accountable for the algorithm outputs.

This is not a cautionary tale from the distant future. It is a composite of real events that have already occurred across financial services, healthcare, hiring, and customer experience. It serves as a preview of what awaits organizations that treat AI governance as a compliance checkbox rather than a strategic imperative.

AI governance is not a technology problem.
It is a leadership problem that belongs in the boardroom.

In this chapter, we will examine what AI governance means for C-suite leaders and board directors, not the technical implementation details that belong to the Chief AI Officer or engineering teams, but the strategic architecture of accountability, oversight, and trust that only leadership can build. We will move from principles to practice, from philosophy to frameworks, and from risk to competitive advantage.

Why Traditional Governance Frameworks Fail AI

Every major corporation has governance structures. Audit committees, risk frameworks, compliance programs, and legal review processes. These are systems that were built over decades to manage familiar categories of organizational risk. They are, in many respects, excellent at what they were designed to do.

AI introduces a fundamentally different class of governance challenge. Traditional governance assumes that a human being made a decision that can be documented, reviewed, and attributed. When a loan officer denies a mortgage application, there is a person, a rationale, a paper trail, and a clear chain of accountability. When an algorithm denies 40,000 mortgage applications, the decision was made by a statistical pattern learned from data, and accountability diffuses across data scientists, product managers, business owners, and the executives who funded the initiative.

The speed problem. AI systems make decisions at a scale and velocity that human oversight cannot match in real time. The fraud detection model makes thousands of decisions per second. Your content recommendation engine determines what 200 million subscribers see within milliseconds of their opening the app. Traditional governance operates at the speed of quarterly reviews and annual audits. AI operates at the speed of light.

The opacity problem. Many high-performing AI models, particularly deep learning systems, are not easily interpreted even by their creators. This is the "black box" problem, and it creates a genuine accountability gap that governance must address structurally.

The drift problem. Unlike a policy document or a process flow, AI models change their behavior over time as they encounter new data. A model that was fair and accurate when deployed may become biased and unreliable eighteen months later without anyone changing a line of code. Governance frameworks built around point-in-time reviews are structurally inadequate for systems that continuously evolve.

I want to pause the drift problem because it is the one that created the Chapter 1 crisis.

Model drift gets attention. Knowledge drift does damage. When a policy changes and the knowledge base is not updated, the AI does not know it is drifting. It continues to retrieve outdated information and deliver it with the same confidence it had for accurate information the day before. There is no error message. There is no degradation in performance metrics. The system looks healthy while quietly delivering wrong answers. Governance frameworks that monitor model performance but not knowledge currency are governing half the system. The half they are lacking is half the customer experience.

The accountability diffusion problem. When an AI system produces harmful output, who is responsible? The data scientist who built it? The product manager who scoped it? The business executive who approved it? The vendor whose foundation model underlies it? In most organizations today, the honest answer is no one, which means no one at all.

Effective AI governance begins with honest acknowledgment that existing frameworks are likely insufficient not because an organization is poorly managed, but because AI represents a genuinely new category of risk that requires new governance architecture.

The Trust Architecture: A Framework for AI Governance

After working across Fortune 100 organizations in financial services, technology, healthcare, and consumer products and having overseen AI governance, I have come to believe that effective AI governance can be understood as a three-layer architecture: the Foundation Layer, the Operating Layer, and the Trust Layer. Each layer has distinct responsibilities, distinct stakeholders, and distinct failure modes.

The Foundation Layer: Values Before Algorithms

Every AI governance framework begins with a question that is deceptively simple and genuinely difficult: What do we believe? Not what is technically possible, not what is currently legal, not what our competitors are doing. What do we, as an organization, prioritize about how AI should treat people?

This is not a technology question. It is an ethics and values question. The question belongs to the most senior leadership of the organization. The answers articulated will cascade through every subsequent governance decision. If your organization has not explicitly answered them, AI systems are implicitly answering them through trained data, optimized metrics, and the accepted failure modes.

The Foundation Layer includes four essential components:

An explicit AI Ethics Statement that goes beyond platitudes. Not 'We are committed to responsible AI' but specific articulations: 'We will not use AI to make final decisions about credit, employment, or healthcare without a human review mechanism. We will not deploy AI systems whose decision-making we cannot explain to affected customers in plain language.'

A Risk Taxonomy that classifies AI use cases by their potential for harm. High-risk applications, those affecting individual rights, safety, financial outcomes, or significant personal decisions require fundamentally different governance than low-risk applications like content personalization or operational efficiency tools.

- A set of Absolute Limits: things an organization will not do with AI, regardless of competitive pressure or short-term financial benefit. These are non-negotiable, and they should be established at the board level, not the product team level.
- A Values-to-Metrics Translation that answers the question: How will we know if we are living our values? Good intentions cannot be audited. Good metrics can.

The Operating Layer: Accountability in Practice

The Operating Layer is where governance meets execution. It encompasses the organizational structures, processes, and systems that ensure your AI values are reflected in AI systems. This is the layer where most governance frameworks live and where most governance frameworks fail because it creates structure without accountability. The four pillars of an effective Operating Layer are:

Pillar 1: Clear Ownership. Every AI system deployed should have an owner—an individual whose performance evaluation is directly tied to the system's responsible operation. This person is not the CTO. They are a business leader with both the authority to act and the accountability to answer for outcomes. In many organizations, this role sits awkwardly between technology and business, owned by neither, accountable to no one. Fix this explicitly.

Pillar 2: A Structured Review Process. Before any AI system is deployed, and at regular intervals thereafter, it should undergo a structured review that examines the data it was trained on and the biases it may contain. What metrics is it optimizing for, and are there cases where optimizing those metrics could produce harmful outcomes? How will it be monitored in production, and what thresholds trigger human review or automatic shutdown? Who is accountable for each of these questions?

Pillar 3: Continuous Monitoring. Deployment is not the end of governance; it is the beginning of the most critical governance phase. AI systems need to be continuously monitored for performance drift, bias emergence, and unintended consequences, not just quarterly. This requires both technical infrastructure (automated monitoring dashboards and anomaly detection) and organizational infrastructure (clear escalation paths, designated responders, and defined action thresholds).

Pillar 4: Incident Response. When—not if—your AI system produces harmful or unexpected output at-scale, how will you respond? Organizations that plan their AI incident response after the incident occurs are organizations that manage crises through press releases rather than through customer protection. Your AI incident response protocol should be as detailed and practiced as your cybersecurity incident response protocol. It should include immediate containment procedures, customer notification frameworks, regulatory disclosure timelines, and root cause analysis processes.

The Trust Layer: Governance as Competitive Advantage

The Trust Layer is where most organizations stop thinking about governance and start thinking about public relations. This is a strategic error. The Trust Layer is not about communication; it is about the external-facing architecture of accountability that allows customers, regulators, and partners to independently verify that AI systems are operating as designed.

This distinction matters. Trust built on communication is fragile; it collapses the moment a crisis reveals the gap between what was said and what was true. Trust built on verifiable accountability is durable; it can survive crises because the structures to detect, disclose, and respond to problems are demonstrably present.

The organizations that will win the AI trust race are not the ones with the best PR teams. They are the ones with the most credible accountability structures.

The Trust Layer has three components: **Transparency Architecture** (what you make visible to external stakeholders and how), **External Accountability Mechanisms** (audit rights, third-party assessments, regulatory cooperation), and **Customer Agency** (the degree to which customers affected by AI decisions can understand, challenge, and appeal those decisions).

In the context of customer agency specifically, organizations serving millions of customers with AI-powered decisions need to grapple with an ethical and strategic question. When an AI system decides what materially affects a customer—denying a service, flagging an account, adjusting a price—what right does that customer have to understand why? What mechanism do they have to challenge the decision? What recourse do they have if the decision was wrong?

These are not hypothetical questions. The European Union (EU) AI Act has made legal requirements for high-risk AI systems across European markets. U.S. regulatory frameworks are developing rapidly. Customers are increasingly demanding them. The organizations that build robust customer agency mechanisms now—not because regulators require them, but because they reflect genuine respect for customer dignity—will be positioned to retain customer trust as these expectations become universal.

The C-Suite Governance Mandate

AI governance cannot be entirely delegated to a Chief AI Officer or a Chief Data Officer. It requires active engagement from every member of the C-suite.

The CEO: Setting the Accountability Culture

The CEO's governance role is primarily cultural. The tone at the top is not a cliché, it is a literal description of the mechanism by which organizational values either constrain or enable AI behavior. When the CEO demonstrates through resource allocation, promotion decisions, and what they celebrate and hold accountable that responsible AI is a genuine priority rather than a communications talking point, the organization responds accordingly. When the CEO signals, consciously or not, that velocity and scale eclipse every other consideration, the governance structures beneath them don't collapse overnight; they erode quietly, decision by decision, until the organization discovers its exposure only after the damage is done.

CEOs should ask themselves regularly: In the last six months, have I made any decision that prioritized caution over speed in an AI deployment context? Have I held anyone accountable for an AI failure in a way that was visible to the organization? Have I explicitly protected someone who raised an AI governance concern, even when that concern slowed a product launch? If the honest answer to all three is no, the governance framework may be more aspirational than operational, which is not uncommon and entirely correctable, but only if leadership chooses to act on it.

The CFO: Funding Governance as Infrastructure

Finance leaders often view AI governance as a cost center: a compliance expense to be minimized rather than an investment to be optimized. This framing is both financially and strategically incorrect. The appropriate analogy is not compliance spending; it is infrastructure spending. You would not build a skyscraper without investing in structural engineering. You should not build an AI-powered enterprise without investing in the governance infrastructure that keeps it structurally sound.

CFOs are uniquely positioned to make the case for AI governance as a financial priority by ensuring it carries a dedicated budget and by building risk models that translate governance failures into the language finance understands regulatory exposure, litigation costs, customer attrition, and reputational impact. When those figures are honestly reflected in the risk-adjusted return on AI investments, governance stops looking like overhead and starts looking like protection.

The Chief Legal and Compliance Officer: From Reactive to Proactive

Legal and compliance functions have historically engaged with AI reactively reviewing contracts with AI vendors: responding to regulatory inquiries and advising on specific high-risk deployments. This posture is no longer sufficient. The legal and regulatory landscape for AI is evolving so rapidly that organizations that engage with it reactively will consistently find themselves behind the compliance curve.

Chief Legal Officers should build proactive regulatory intelligence capabilities, systematically tracking AI regulatory development across all relevant jurisdictions, engaging directly with regulators (not just reactive responses to enforcement), and participating in industry standards bodies shaping the governance landscape. The organizations that help write the rules are better positioned than those who merely comply with them.

The Chief Human Resources Officer:
The Human Side of the Algorithm

AI governance has a workforce dimension that is frequently overlooked in technology-focused governance conversations. AI systems are designed, built, deployed, and monitored by human beings who operate within organizational cultures that have their own values, incentives, and blind spots. A governance framework that focuses exclusively on technical systems while ignoring the human systems that produce them is fundamentally incomplete.

CHROs should ensure that AI governance responsibilities are explicitly embedded in job descriptions and performance evaluations, that employees have clear and protected channels for raising AI concerns, and that the organization's approach to AI does not create perverse incentives that reward speed over care and celebrate scale without accountability.

The Chief Customer Officer or CXO:
The Customer Accountability Function

If an organization has a Chief Customer Officer or equivalent, that leader has perhaps the most direct AI governance accountability in the C-suite: ensuring that AI systems deployed in customer-facing contexts serve customer interests, not just business metrics that correlate with customer interests in normal conditions but diverge from them under stress.

Customer-facing AI systems are where trust is won and lost. Recommendation engines that serve customer interests build loyalty. Recommendation engines that maximize short-term engagement metrics while undermining customer well-being erode it, often invisibly, until customers stop trusting a brand entirely. The CXO should be the internal advocate for customer interests in every AI deployment decision, with enough organizational authority to slow or stop deployment when those interests are inadequately protected.

The Board's AI Governance Mandate

Board oversight of AI is one of the most consequential and most underdeveloped areas of modern corporate governance. A 2024 survey found that fewer than 30% of Fortune 500 boards had established explicit AI oversight frameworks, and only 12% had a director with genuine AI domain expertise. Yet these same boards approve AI investment strategies that will determine their companies' competitive positions for the next decade and their regulatory exposure for the next generation.

This gap between AI's strategic importance and boards' governance sophistication is not sustainable. Regulators are noticing it. Institutional investors are noticing it. The costs of the gap are beginning to manifest in real organizational crises.

What the Board Should Oversee (And What It Should Not)

Boards frequently make one of two errors in AI governance: they either delegate all oversight to management (abdicating their responsibility) or they attempt to govern at an operational level for which they lack information and expertise (overstepping their role). The board's proper role in AI governance sits between these extremes.

The board should oversee strategy, risk tolerance, and the accountability culture and hold management accountable for operational governance. It should not attempt to audit individual AI systems, review technical documentation, or second-guess product decisions. The board's leverage is not operational; it is structural. It sets the conditions under which operational governance either thrives or fails. Specifically, boards should:

- Approve the organization's overall AI governance framework and risk appetite statement, ensuring it reflects genuine values rather than generic platitudes.
- Receive regular reporting on AI risk not just AI opportunity with enough specificity to identify governance gaps and accountability failures.
- Ensure that someone at the management level is clearly accountable for AI governance outcomes, with the authority to act and the responsibility to answer for them.
- Oversee the organization's engagement with AI regulation, including its posture toward developing

regulatory frameworks and its readiness for likely future requirements.

- Commission periodic independent assessments of the organization's AI governance maturity, not relying solely on internal reporting.

Building an AI-Literate Board

Effective AI oversight requires some baseline level of AI literacy among board members. It means every director engages with questions about how AI systems are built, what can go wrong, and what meaningful oversight looks like.

Organizations serious about AI governance should invest in director education not through vendor-sponsored conferences with glossy presentations, but through substantive engagement with AI practitioners, ethicists, and regulators who can provide honest assessments of where risks lie. Boards should also consider whether their membership includes individuals with genuine AI domain expertise, whether through permanent appointments or advisory relationships.

The Questions Every Board Should Be Asking

Based on my experience overseeing AI governance, I have developed a core set of questions that every board should be asking management at least annually—and in some cases, quarterly.

Accountability:

- Who is named as accountable for AI governance outcomes, and how is that accountability reflected in their performance evaluation?
- What happened the last time an AI system produced an unexpected or harmful outcome? How was it detected? How was it addressed?

Risk:

- What are the three AI-related risks that keep our Chief AI Officer or Chief Risk Officer up at night? What is being done about them?
- Are we using AI in any context that regulators in our key markets would classify as high risk under current or anticipated regulations? What is our compliance posture?

Customer Trust:

- Do our customers have the meaningful ability to understand, challenge, and appeal AI-driven decisions that affect them? How do we know?
- Have our AI systems produced any outcomes in the past twelve months that we would be uncomfortable explaining to our customers, regulators, or the press?

Competitive Positioning:

- Is our AI governance posture a competitive advantage, a competitive disadvantage, or a neutral factor in our industry? What evidence supports that assessment?
- What governance capabilities are we building now that will differentiate in three to five years?

The Regulatory Landscape: From Compliance to Strategy

The AI regulatory environment is evolving at an unusual pace and with unusual reach. Unlike previous technology regulatory waves which developed over years and often reached meaningful enforcement only after a decade or more, AI regulation is moving quickly, and early indications suggest regulators intend to enforce seriously.

The EU AI Act, which entered into force in 2024 and is rolling out over a multi-year implementation schedule, is the most comprehensive AI regulation yet enacted. It classifies AI systems by risk level, imposes substantial requirements on high-risk applications, and creates compliance and audit infrastructure. The Act's extraterritorial reach applies to any AI system whose outputs are used in the EU, regardless of where the system was developed and makes it an effective global compliance floor for large organizations.

In the United States, federal AI regulation has developed more slowly, but sector-specific regulatory guidance is proliferating rapidly across financial services (through the FTC, OCC, and CFPB), healthcare (through the FDA and HHS), and employment (through the EEOC). State-level AI legislation, led by Colorado and California, is filling the federal void. The patchwork nature of

U.S. AI regulation creates its own compliance challenges and opportunities for organizations that engage proactively with the regulatory development process.

The strategic insight here is critical and counterintuitive: regulatory compliance, pursued minimally and reactively, is a cost center. Regulatory engagement, pursued proactively and strategically, is a competitive advantage. Organizations that participate in shaping AI governance standards through industry associations, regulatory comment processes, and direct engagement with policymakers have influence over the frameworks they must comply with. They have advanced visibility into regulatory direction that informs product and investment decisions. They develop internal governance capabilities that make compliance straightforward rather than expensive.

The organizations that help write the rules of AI governance will be better positioned than those that merely comply with them.

From Policy to Practice: Building A Governance Roadmap

Governance frameworks have a notorious failure mode: they produce excellent documents that have no operational effect. The AI Policy sits on the intranet. The Ethics Committee meets quarterly and issues guidance that product teams note and ignore. The risk framework is updated annually and never consulted in real decisions.

Avoiding this failure mode requires deliberate design of the interface between governance structures and operational decisions.

Governance must be embedded in the processes through which AI systems are built, deployed, and monitored, not maintained as a parallel structure operating on a different timeline and at a different organizational level.

A 90-Day Quick-Start for C-Suite Leaders

If you are a C-suite leader reading this chapter and wondering where to start, here is a practical 90-day roadmap:

Days 1–30: Assess and Illuminate

- Commission a complete inventory of all AI systems currently deployed in your organization, including third-party AI embedded in vendor products.
- For each system, document: what decisions it makes or influences, who owns it, how it is monitored, and the process for escalating concerns.
- Identify three highest-risk AI deployments and assign an owner to each with explicit accountability.

Days 31–60: Establish Structure

- Appoint or designate an AI governance function even if this is initially a part-time responsibility for an existing leader with clear authority and C-suite access.
- Draft an AI Risk Taxonomy that classifies your current use cases by risk level and identifies the governance requirements for each level.

- Establish a cross-functional AI Governance Council with representatives from Legal, Risk, Technology, Product, and at least one business line. Define its mandate, meeting cadence, and decision-making authority.

Days 61–90: Build Accountability Infrastructure

- Develop a standardized AI Review Template for new deployments with a minimum set of questions that must be answered before any AI system reaches production.
- Establish monitoring dashboards for your highest-risk AI systems with defined thresholds for human review.
- Prepare your first AI Governance Report for the board, establishing the reporting cadence and framework going forward.

The Governance Maturity Model

Organizations progress through predictable stages of AI governance maturity. Understanding where your organization sits on this spectrum and what the next stage looks like is essential for building a credible roadmap.

Stage 1 – Ad Hoc: No systematic governance. AI decisions are made by individual teams. Risk is managed informally. Most large organizations currently operate at this stage for at least some of their AI portfolio.

Stage 2 – Defined: Basic policies and processes are in place. There is a designated owner for AI governance. High-risk deployments receive structured review. Monitoring exists but may be inconsistent.

Stage 3 – Managed: Governance is embedded in development and deployment processes. Monitoring is systematic. Accountability is clear and reflected in performance management. The board receives regular, substantive AI risk reporting.

Stage 4 – Optimizing: Governance drives competitive advantage. The organization actively engages in external governance development. Customer-facing accountability mechanisms are a differentiator. AI governance quality is measurable and improving.

Most Fortune 500 organizations currently operate between Stages 1 and 2. The organizations that will capture the AI trust premium over the next decade are those that move deliberately to Stages 3 and 4, not because regulators require it, but because customers demand it and the competitive logic is overwhelming.

The Trust Dividend: Why Governance Is a Growth Strategy

I am closing this chapter by challenging the idea that governance and growth conflict with one another. This framing, which is implicit in the way most organizations approach AI governance, is both empirically wrong and strategically dangerous. The empirical case: Research consistently shows that consumer trust is among the most powerful drivers of brand preference, customer lifetime value, and willingness to share data. Data sharing is the fuel of AI advantage; organizations with more and better data build better models. Organizations that build and maintain deep customer trust are not sacrificing AI capability for governance compliance, they are building the data foundation that enables AI capability.

The strategic case: The AI trust gap between leading and lagging organizations is widening, not narrowing. Early movers in AI governance are establishing durable competitive positions through regulatory relationships, customer data advantages, talent attraction, and crisis resilience that late movers will find increasingly difficult to close. The cost of building governance infrastructure early is substantially lower than the cost of building it reactively after a major governance failure.

I call this the trust dividend: the measurable, sustainable competitive return on investment in AI governance that manifests across customer retention, regulatory efficiency, talent quality, and brand premium. The trust dividend is not a soft benefit that sounds good in an annual report. It is a financial reality that increasingly shows up in customer acquisition costs, customer lifetime value, regulatory compliance costs, and, for public companies, valuation multiples.

The question is no longer whether your organization will invest in AI governance. The question is whether you will invest proactively and capture the trust dividend or reactively, after a governance failure has already extracted its price. The leaders who grasp this dynamic early who see AI governance not as a constraint on AI ambition but as the foundation on which sustainable AI advantage is built will be the ones who look back in a decade and understand that responsible AI was not the careful path. It was the smart path.

A Final Note: The Algorithm Has No Values. You Do.

It's easy to feel like AI is moving faster than any of us can steer it, and that the best we can do is respond to whatever it produces.

Algorithms do not have values. They do not have intentions. They do not care about your customers, your brand, or your organization's long-term interests. It optimizes the given objective with the data it was trained on, within the designed constraints. Every input is a human decision.

The governance challenge of the AI era is not managing technology; it is managing the impact of technology. It is ensuring that human values specifically, an organization's values about how it treats the people it serves are accurately and completely reflected in the systems deployed at-scale.

The trust algorithm, at its core, is not an algorithm at all. It is a set of choices about accountability, transparency, privacy, decisions, and loyalty. Those choices are yours to make.

Make the choices deliberately.

11. THE TRANSFORMATION LEADER'S PLAYBOOK

By now, the message is unmistakable: Generative AI is not another technology cycle to "monitor." It is the inflection point for reengineering how organizations deliver service, how employees do high-value work, and most critically how trust is earned at-scale. The leaders who treat Gen AI as a tool upgrade will get incremental automation. The leaders who treat it as an operating model shift will build structural advantage: lower cost-to-serve without degrading experience, faster decision velocity without sacrificing control, and a customer relationship that becomes more resilient over time rather than more fragile.

The hard part for COOs, CCOs, and transformation leaders is not believing in the potential. It is deciding where to begin when the constraints are real: customers expect immediate outcomes, boards demand measurable ROI and risk clarity, and employees worry that "efficiency" is code for displacement. In this environment, a misstep is expensive in more than dollars, it can create reputation damage, internal resistance, and a long tail of distrust that slows every subsequent initiative. The organizations that win are the ones that approach Gen AI with discipline: a human-centered roadmap that builds confidence, capability, and control in parallel.

That is why I anchor Gen AI programs in a simple but rigorous five-stage roadmap—**Vision → Pilot → Scale → Govern → Evolve**—designed not just to deploy AI, but to deploy it in a way that earns durable trust. Sustainable success rarely comes from a single "big bet" or a well-timed platform purchase. It comes from sequencing decisions correctly: aligning outcomes before technology, learning before expansion, operationalizing before celebrating, and building governance as an enabler rather than a brake.

Stage 1: Vision—Begin with the Why: The most common failure pattern is starting with a demo and working backward into a strategy. It feels efficient, but it quietly creates misalignment: teams optimize for what the tool can do rather than what the business must accomplish. A credible vision defines the outcomes that matter reducing repeat contacts, improving first-contact resolution, shrinking time-to-resolution, strengthening compliance, or increasing loyalty and connecting them directly to the brand promise. It also names the non-negotiables upfront: which customer moments must remain human-led, what "good" looks like from a trust perspective, and where risk tolerance is low. Leaders who do this well don't just write a strategy, they create decision filters that prevent scope creep, vendor-driven detours, and KPI theater.

Stage 2: Pilot—Start small, learn quickly: Pilots are not miniature deployments; they are learning engines. The goal is to validate value and surface friction early: how customers respond, how frontline teams adopt, where the model fails, what data gaps exist, and what governance requirements will matter at-scale. The strongest pilots are tightly scoped, high-frequency use cases where iteration speed is possible, and failure is containable. They produce metrics but they also produce something more important: organizational belief. Skeptics become curious. Middle managers see how work changes. Frontline employees experience augmentation rather than abstraction. In practice, that "confidence flywheel" becomes the difference between a pilot that stays trapped in PowerPoint and one that earns permission to scale.

Stage 3: Scale—Build something that lasts: A successful pilot proves the possibility; scaling creates permanence. This is the stage that separates organizations that "use AI" from organizations that operate with AI. Scaling requires enterprise-grade architecture, integration into core workflows, and process redesign that makes augmentation the default way of working not an optional extra tool. It also requires workforce intent: role clarity, skills pathways, and adoption mechanics. Employees don't resist technology; they resist ambiguity. When people understand how their role changes, how quality is protected, and what growth looks like for them, adoption accelerates and performance stabilizes. Communication, training, and change management are not soft elements: they are operational controls that determine whether value compounds or stalls.

Stage 4: Govern—Protect what you've built: Trust is not recoverable on demand. When it breaks, customers don't file a complaint about your governance model; they simply leave, and regulators take an interest. Governance must be built into the product, not layered on afterward. That means preventing predictable failure modes—hallucinations, bias, inconsistency, data leakage before they reach the customer; establishing clear audit trails and accountability for AI outputs; and embedding human oversight where the stakes require it. It also means being transparent with customers about where AI operates and how data is protected, because the absence of clarity becomes a trust liability. Governance is not a constraint on innovation; it is the condition that allows innovation to scale safely.

Stage 5: Evolve—Stay in motion: Gen AI is not a destination; it is an accelerating capability curve. Models improve, regulations shift, customer expectations rise, and yesterday's differentiation becomes today's table stakes. The organizations that thrive treat deployment as the start of a learning system: continuous feedback loops from customers and employees, routine evaluation of model performance and drift, ongoing retraining and policy updates, and a cadence for experimentation that is both safe and fast. The goal is not to "implement AI." The goal is to build an organization that can adapt as AI evolves without having to reboot the program every time the technology changes.

Transformation leaders must align functions, establish the operating cadence, and close the gap between strategy and execution with disciplined sequencing. Gen AI will not succeed because it has already been deployed. It will succeed because leaders owned the outcomes, designed for adoption, and protected trust while moving at speed. To make the roadmap tangible, consider how it plays out in practice.

In Stage 1, a global retailer sees customer service costs rising while satisfaction is flat. Instead of starting with vendors, the CCO convenes a cross-functional working session to define the business problem and agree on priorities: reduce repeat contacts, improve first-contact resolution, and protect the brand experience during peak seasons. Those priorities become the governance and investment filter for every Gen AI decision that follows. Technology enters second, not first.

In Stage 2, the organization resists the temptation to deploy across every channel and launches one contained use case: AI-assisted responses for order status inquiries in a single market. Handle time improves, and containment rises, but the more valuable outcome is cultural: frontline associates move from skepticism to advocacy, and leaders see proof without exposing the entire operation to disruption.

In Stage 3, scaling is deliberate: AI is integrated directly into the agent desktop workflow rather than added as a separate tool; process redesign occurs before go-live; and every frontline team receives role-specific enablement that answers the question employees ask: "What does this mean for me?" Adoption accelerates because people were brought along, not left behind.

In Stage 4, governance proves its value: a quality review surfaces a pattern of technically accurate but contextually misleading responses in a specific billing category. Because audit trails and monitoring are in place, the issue is identified quickly, a human review threshold is set for that category, and customers never experience failures at-scale; the program is protected without slowing to a crawl.

By Stage 5, six to twelve months later, AI fluency is embedded not only in service but in workforce planning and proactive outreach. When a new capability emerges, predicting customer frustration before a contact is initiated, the organization doesn't start over. It moves quickly because the infrastructure, governance, talent, and confidence are already in place, while competitors are still stuck proving pilots.

Ultimately, the leaders who win in Gen AI are not the ones who move the fastest in a straight line. They are the ones who build momentum without breaking trust, sequencing ambition with discipline, scaling value while protecting customers, and evolving the operating model so the organization gets stronger with every iteration.

12. STRATEGIC FRAMEWORKS

The gap between Gen AI programs that become enterprise capabilities and those that remain expensive experiments is rarely about model quality. It is about operating discipline. Successful transformations are guided by frameworks that force clarity on priorities, concentrate resources on the few use cases that matter, and create repeatable execution patterns that survive leadership changes, budget cycles, and the inevitable "pilot fatigue." In this chapter, I translate research and operating lessons from leading strategy firms and business publications into a set of practical frameworks CX leaders can use to decide what to build, how to sequence it, and how to scale it without eroding trust.

McKinsey's "Rewired" view is particularly useful because it reframes AI as a full-system transformation rather than a technology project. Across hundreds of at-scale implementations, the organizations that create material enterprise-wide value do six things well at the same time: they anchor Gen AI to strategy, build talent deliberately, redesign the operating model, modernize technology foundations, treat data as a governed asset, and industrialize adoption and scaling. The strategic dimension is where most programs quietly fail. Leaders are often able to describe what they want the model to do, but not why it matters for competitive advantage. A simple test is whether you can explain the initiative without using the term "AI." When the narrative is grounded in outcomes—higher first-contact resolution, fewer repeat contacts, faster cycle times, increased retention, reduced risk exposure, resource allocation becomes easier, and governance becomes less political because the trade-offs are explicit.

The talent dimension is routinely misunderstood as "hire a few data scientists," when the real requirement is role evolution at-scale. Gen AI changes move faster than org charts, so leaders need explicit pathways that help people see a future with AI rather than one replaced by AI. High-performing organizations reskill existing teams into AI-adjacent roles such as prompt and workflow designers, knowledge curators, quality evaluators, and model-risk partners while protecting institutional knowledge that cannot be recreated from documentation.

Expertise distribution is uneven by age and experience, which creates an opportunity to pair digitally fluent employees with seasoned operators who understand customer nuance, policy intent, and edge cases. That pairing accelerates adoption, reduces risk, and prevents the common failure mode where "AI people" build something that works in theory but breaks in the messy reality of customer contact.

The operating model dimension is the difference between speed and stalemate. Traditional hierarchical decision-making cannot keep pace with Gen AI iteration cycles, where learning happens in weeks, not quarters. Organizations that scale create agile, cross-functional product teams accountable for end-to-end outcomes from customer impact to risk posture, so decisions sit close to where the work and feedback occur. In these models, "failure" is treated as managed learning: bounded experiments, controlled exposure, and rapid rollback mechanisms. Without that psychological safety and operating cadence, teams revert to over-engineering and slow approvals, and the program dies by process even when the technology is ready.

Technology and data are often discussed as separate pillars, but in practice, they form a single system. Enterprise value requires platforms that enable secure integration into core workflows, API connectivity across systems of record, scalable compute, and monitoring that detects drift, bias, and quality degradation before customers do. Treating Gen AI as a departmental add-on guarantees fragmentation, inconsistent customer experiences, and weak governance. Data governance is the make-or-break accelerator here, not because leaders need perfect data to start, but because they need clear ownership, quality standards, ethical guidelines, and accessible repositories that enable continuous improvement rather than sporadic change. A less obvious lesson from large enterprises is that the biggest barriers are often cultural, not technical: organizations struggle not because they cannot store data, but because they cannot agree on definitions, accountability, and decision rights especially across functions.

The dimension adoption and scaling is where value is either captured or lost. Gen AI only creates ROI when it changes daily work, not when it lives in a separate tool that employees "should use." Successful programs embed AI into the agent desktop, the knowledge system, the QA workflow, and the operating cadence then tie usage to business outcomes with measurable KPIs and aligned incentives. At maturity, the organization becomes operationally dependent on the capability in the same way it depends on workforce management or CRM: removing it would break throughput, consistency, or quality. That is the real signal when you have scaled AI; it is no longer a project, it is infrastructure.

BCG's three-phase transformation framework complements McKinsey by adding sequencing discipline. In the first phase (roughly months 1–6), leaders focus on short-term wins that are visible, measurable, and low-complexity, proof points that create political capital and fund the next wave.

The nuance is that "quick wins" should not be chosen solely for ease; they should be chosen because they also lay the foundations for instrumentation, data pipelines, evaluation methods, and governance patterns that can be reused. In the second phase (months 6–18), the focus shifts to replication and capability-building, scaling the winning use case across teams and channels while building a repeatable playbook, training, and governance model that enables subsequent deployments to be faster and cheaper. In the third phase (18+ months), the core objective is cultural and structural sustainability: embedding AI literacy broadly, rewarding disciplined experimentation, and building feedback loops that keep the system improving. Without this third phase, organizations often experience a predictable stall: early deployments succeed, but adoption plateaus because the organization never made the behavioral shift from "initiative" to "new way of operating."

Harvard Business Review's cost–knowledge matrix provides a pragmatic decision-making lens for CX leaders to prioritize use cases without being seduced by novelty. It asks two questions that boards and regulators implicitly care about anyway: what is the cost of a mistake, and what type of knowledge is required to do the work well?

Where error costs are high—financial harm, reputation damage, safety, compliance exposure, autonomy must be constrained, and human oversight must be designed in, not bolted on. Where work relies on explicit knowledge structured policies, known procedures, searchable information, Gen AI tends to perform well and can be deployed with confidence. Where work depends on judgment, reading emotional subtext, ethical reasoning, trust repair in ambiguous moments, Gen AI can assist, but it should not lead.

This matrix is especially important because customer-facing

service is deceptively complex: edge cases are the norm, not the exception, and small errors compound into distrust quickly. Many organizations find faster, safer value by deploying agentic capabilities internally first such as knowledge maintenance, case summarization, quality evaluation, and workforce planning insights until governance and reliability are proven. BCG's "head, heart, hands" leadership model is the human operating system that makes these frameworks executable.

"Head" is strategic clarity and systems thinking, articulating a compelling why, connecting initiatives to enterprise outcomes, and making principled choices about what not to do.

"Heart" is trusting leadership: acknowledging employee fear, communicating transparently about role evolution, and creating psychological safety so teams surface issues early rather than hiding them.

"Hands" is disciplined execution: rigorous program management, metrics that measure what matters, obstacle removal, and sustained momentum through the difficult middle period when novelty fades and real change begin. Leaders who over-index on any one dimension create predictable failure modes—vision without empathy generates resistance, empathy without execution produces stalled value, and execution without vision optimizes the wrong outcomes at-scale.

Placed together, these frameworks form an integrated approach that CX leaders can run like an operating cadence. McKinsey provides the capability architecture: the six dimensions you must build across to generate enterprise value. BCG provides the sequence: how to prove value, scale responsibly, and institutionalize the change. HBR provides prioritization guidance on where to deploy Gen AI based on error cost and knowledge complexity. And the head–heart–hands model provides the leadership behavior required to sustain performance as the organization absorbs a fundamentally new way of working.

Frameworks do not guarantee success, but they dramatically improve the odds by removing the two most common causes of failure: scattered investments and unmanaged trust risk. The winners in the Gen AI era will not be the organizations that invent brand-new playbooks. They will be the ones to apply battle-tested frameworks with focus and discipline, treating Gen AI as an operating model transformation, not a technology experiment, and to build value in a way that customers, employees, and boards can trust.

13. SCALING TRUST IN A REGULATED WORLD

Every transformation leader working in a regulated industry has heard the same objection: *"That won't work here—we're too regulated."* On the surface, the concern is understandable. Financial institutions, healthcare providers, insurers, and government agencies operate in environments where the consequences of failure are high.

These organizations manage people's savings, medical information, legal rights, and, in some cases, their safety. Regulators expect precision, accountability, and documentation at every step. A single breakdown in the process can trigger fines, litigation, reputation damage, or regulatory intervention that lingers for years. In these environments, caution is not resistance to innovation; it is a professional obligation.

Yet that same sense of responsibility is precisely why these industries cannot afford to remain static. The sectors most hesitant to adopt generative AI are often the ones best positioned to benefit from it. In regulated markets, the ultimate differentiation is not convenience alone but trust: the assurance that decisions are consistent, information is accurate, and customer interests are protected.

When designed with the rigor these sectors demand, Gen AI does not undermine trust; it can strengthen the very systems that sustain it. The real opportunity is not automation for its own sake, but the creation of systems that are more consistent, transparent, and auditable than purely manual processes have ever been.

This tension creates what might be called the compliance paradox. For decades, organizations have attempted to reduce regulatory exposure by layering additional controls into the customer experience—scripts, disclosures, approval chains, documentation requirements, and escalating levels of oversight. While well-intentioned, these measures often produce unintended consequences. Service becomes slower, employees struggle to navigate complex requirements in real time, and customers feel overwhelmed by legal language that obscures rather than clarifies the intent of the interaction. In trying to protect trust, organizations sometimes create experiences that erode it.

Generative AI offers a path to resolve this paradox. Rather than treating compliance as a friction layered on top of service, AI allows it to become embedded within the service itself. When implemented thoughtfully, AI can improve regulatory compliance consistency, reduce operational risk, and make compliance more visible, thereby strengthening confidence among customers, regulators, and employees alike.

One of the immediate advantages appears in disclosure and policy adherence. Many compliance failures do not stem from negligence but from the realities of human cognition, associates under pressure managing multiple systems, policies, and customer questions simultaneously. AI systems can automatically insert required disclosures or policy guidance at the exact moment they are needed within an interaction, ensuring that regulatory requirements are delivered accurately and consistently.

Now imagine the Chapter 1 scenario unfolding in a regulated

environment. A banking customer asks about the terms of a financial product. The AI uses outdated terms that refer to a fee structure that was changed three months ago to comply with new regulatory guidance. The customer calls to confirm. The agent, consulting the same uncorrected knowledge base, confirms the outdated terms.

The customer makes a financial decision based on those terms. When the actual terms apply, the customer has been materially misled not by a rogue employee, not by a malicious system, but by a knowledge management failure that cascaded across both channels.

In this scenario, the remediation is not just an apology. It is a potential regulatory finding, a compliance event that could trigger enforcement action and reputational harm far beyond the individual customer. This is why Compliance by Design must include knowledge governance as a first-order priority. In regulated industries knowledge accuracy is not a service quality issue; it is a compliance obligation.

The next big change is about accountability: being able to see inside AI decisions and explain them. In traditional operations, reconstructing an interaction for regulatory review can require manual investigation across disparate systems. AI-enabled service platforms can automatically capture, transcribe, and structure interactions, preserving both the content of the exchange and the operational context surrounding it. The result is not merely documentation but demonstrable accountability. Regulators increasingly seek evidence that organizations can show how decisions were made and what safeguards were in place at the time, capabilities that AI systems are uniquely positioned to provide.

Generative AI can also serve as a real-time safeguard during live interactions. Advanced monitoring capabilities allow systems to detect language or behaviors that could create regulatory exposure while a conversation is still unfolding. Instead of discovering problems during post-interaction audits, organizations can intervene immediately, guiding employees toward compliant alternatives or appropriately escalating complex situations. In effect, AI functions as a continuous compliance partner, supporting employees in moments where judgment and accuracy matter.

Bias detection represents another area where AI, when properly governed, can strengthen oversight. Patterns that might take months or years to detect through manual analysis: disparities in loan approvals, claim processing decisions, or service outcomes across demographic groups can be surfaced quickly through systematic monitoring. At the same time, leaders must recognize that AI systems themselves inherit the characteristics of the data used to train them.

Responsible governance, therefore, begins upstream, ensuring that training data, evaluation criteria, and model design do not embed discriminatory assumptions. Bias introduced at the system level inevitably manifests in customer outcomes, often in subtle ways that are difficult to identify once deployed. Addressing this risk requires both technical safeguards and a broader ethical commitment to fairness.

Compliance itself can also become more dynamic through AI-enabled learning systems. Traditional compliance training often occurs in isolated annual sessions that employees complete once and rarely revisit. Generative AI enables organizations to shift toward continuous, context-aware learning embedded in the flow of work. When policies evolve or regulatory guidance changes, employees receive immediate guidance tied directly to the situations they encounter. Instead of compliance being a periodic event, it becomes an ongoing practice—reinforced through daily interaction rather than episodic instruction.

Within executive discussions, these opportunities frequently collide with two competing instincts. The first is the instinct to minimize risk by delaying adoption until the regulatory environment becomes clearer. The second is the recognition that customer expectations continue to evolve regardless of an organization's pace of change. Personalization, responsiveness, and clarity are becoming baseline expectations even in highly regulated industries. Competitors are experimenting with new capabilities, and regulators themselves are increasingly evaluating whether AI, when governed responsibly, can produce more consistent outcomes than purely human-driven processes.

The organizations that navigate this tension successfully tend to adopt Compliance by Design philosophy. Rather than treating governance as a control applied after technology is built, they incorporate regulatory considerations into the system's architecture from the beginning. Governance frameworks, human oversight mechanisms, and audit capabilities are designed as integral components of the solution rather than retrofitted later in response to risk concerns.

This approach requires building robust oversight structures that clearly define when human judgment must remain in control and when AI can safely augment or automate processes. Transparency with customers has become equally important. Individuals deserve clarity about when AI is part of their experience and how their information is being used.

Continuous monitoring is another essential element. AI systems are not static assets; they evolve as data, models, and regulatory expectations change. Organizations must regularly evaluate system performance, identify model drift, and retrain models when necessary. Far from being an extraordinary measure, this ongoing attention reflects sound operational discipline in an environment where both technology and regulation continue to advance.

The final element of Compliance by Design involves a shift in how organizations engage with regulators. The most forward-looking institutions treat regulators not as obstacles but as stakeholders in the development of responsible AI practices. By sharing governance approaches, demonstrating safeguards, and participating in shaping emerging standards, companies can build constructive relationships that reduce uncertainty and accelerate responsible adoption.

Real-world examples illustrate how this approach can reshape both compliance and customer experience. In banking, generative AI can translate complex policy language or lending decisions into clear, accessible explanations that customers understand. When transparency improves, confusion declines, and complaint volumes decline. In healthcare, AI can assist patients with scheduling, identity verification, and eligibility checks while remaining carefully constrained from providing clinical advice. The result is a faster and more consistent experience that reduces operational strain while preserving the human expertise required for sensitive medical decisions.

Organizations that embed Compliance by Design from the outset through legal-approved knowledge bases, auditable interaction records, rigorous testing, and structured oversight often discover unexpected outcomes. Rather than slowing innovation, strong governance accelerates it. When regulators, employees, and customers trust the system's foundation, organizations gain the confidence to scale new capabilities more quickly because risks are understood and managed.

Compliance by Design is not about slowing AI down—it is about creating the structural integrity required to go further.

Regulated industries have spent decades building credibility with the customers and institutions they serve. That credibility is not an obstacle to AI adoption; it is the very asset that makes responsible adoption meaningful. Leaders who approach generative AI with that perspective will do more than avoid regulatory pitfalls. They will create organizations where technology strengthens accountability, service becomes clearer and more consistent, and trust earned carefully over time becomes a competitive advantage that others struggle to replicate.

14. THE GEN AI WORKFORCE OF THE FUTURE

Whenever a breakthrough in technology enters the workplace, the first question people ask is almost always the same: *What will happen to jobs?* Each technological wave ranging from automation in manufacturing to digital transformation in banking has triggered some version of this concern. The arrival of generative AI has dramatically amplified it. Headlines frequently frame the technology as a replacement engine, predicting large-scale layoffs and a future in which algorithms handle nearly every interaction once performed by people.

That narrative, while attention-grabbing, overlooks a more complex and realistic transformation already unfolding. The true impact of generative AI is not the removal of humans from work but the redefinition of how human capability is applied. In customer-facing environments, Gen AI changes the balance between routine execution and high-value judgment. Tasks that once consumed enormous amounts of employee time—searching knowledge bases, documenting interactions, navigating fragmented systems, or repeating standard information—can now be automated or augmented. What remains is the work that requires uniquely human strengths: empathy, judgment, relationship-building, and problem resolution in moments that matter.

In many ways, generative AI is less about workforce reduction and more about workforce elevation.

In this environment, AI becomes an always-present knowledge

partner. It surfaces relevant policies, past interactions, product details, and recommended responses instantly, eliminating the need for associates to search through fragmented systems or rely solely on memory. Instead of spending valuable minutes navigating databases, employees receive context-aware insights directly within the conversation workflow.

AI can also provide behavioral guidance that improves the emotional quality of interactions. For example, systems can recognize signals of frustration or confusion in customer language and prompt agents with suggestions such as acknowledging concerns, clarifying expectations, or offering alternative solutions. These prompts do not replace empathy; they reinforce it by ensuring employees have the situational awareness needed to respond thoughtfully.

By removing the constant cognitive strain associated with system navigation and documentation, AI frees employees to focus on the most important aspect of service: human connection. Conversations become less transactional and more relational. Associates can listen carefully, think creatively about solutions, and engage customers with genuine attention rather than procedural urgency.

The result is not only operational improvement, faster resolutions, and higher productivity, but also healthier employee experiences. When the work becomes less about surviving the system and more about solving real problems, engagement rises and burnout declines.

A New Workforce Architecture

As generative AI becomes embedded in service organizations, the workforce itself will evolve. Five years from now, the most advanced customer experience organizations will likely look very different from the contact centers of the past.

Frontline associates will operate as AI-augmented advisors capable of handling complex customer needs with both speed and emotional intelligence. Quality and operations teams will shift from monitoring agent performance alone to continuously refining the AI systems that support those agents. These teams will analyze patterns, identify opportunities for improvement, and ensure that AI-generated guidance aligns with both company policy and customer expectations.

New orchestration roles will emerge to ensure that digital channels, automated systems, and human support work together seamlessly. Rather than viewing automation and human service as separate layers, these professionals will design end-to-end customer journeys where each interaction, whether handled by AI or a person, flows naturally into the next.

Data specialists will become central to service operations as well. Their responsibility will be to ensure that the knowledge feeding AI systems remains accurate, current, and compliant with evolving regulatory requirements. This work will involve maintaining knowledge repositories, validating training data, and monitoring system performance to ensure that recommendations remain trustworthy.

Leadership roles will also evolve. Instead of measuring success primarily through volume-based metrics such as handle time or case closure rates, executives will increasingly focus on indicators of long-term value: customer trust, relationship strength, and the ability of service interactions to influence loyalty and growth.

In this emerging model, the workforce is more specialized, capable, and strategically aligned with the organization's purpose.

The Imperative of Reskilling

For leaders, the defining challenge of the Gen AI era is workforce development at scale. The transition required is less about learning new software and more about evolving the way work itself is performed.

Employees who followed scripts must now develop confidence to engage customers with empathy and judgment. Process-oriented roles must expand to include problem-solving and decision-making. Associates who previously spent most of their time navigating systems will increasingly act as relationship stewards, guiding customers through complex situations.

This shift demands a new approach to workforce development. Traditional training models, short on-boarding sessions or annual compliance modules—are insufficient for preparing employees for AI-augmented work. Organizations must instead create continuous learning environments where employees build capabilities over time.

These capabilities include emotional intelligence, adaptability, critical thinking, and ethical judgment. While often categorized as "soft skills," they become the most strategically important skills in an AI-enabled organization. As technology handles routine cognitive tasks, the value of human judgment and relational intelligence increases significantly.

Leaders must therefore view reskilling not as a defensive exercise but as a strategic investment. The organizations that commit early to developing these capabilities will create a uniquely equipped workforce to thrive alongside AI.

Navigating Employee Sentiment

Despite these opportunities, it would be unrealistic to assume that employees universally welcome the arrival of AI. Many workers worry that their roles may disappear or that they will struggle to keep pace with technological change. Others simply feel fatigued by the constant evolution of systems and expectations in modern workplaces.

Addressing these concerns requires transparency and empathy from leadership. Avoiding the conversation only deepens mistrust. Leaders must acknowledge employees' fears openly while providing a clear, credible vision of how AI will enhance, rather than diminish, people's roles within the organization.

Trust begins internally. An employee who feels threatened or insecure cannot effectively communicate reassurance and confidence to customers. Conversely, when employees feel supported and empowered, confidence becomes visible in every interaction.

Organizations that navigate this transition successfully tend to do several things well. They create clear pathways for reskilling so employees understand how their careers can evolve. They recognize and reward individuals who embrace AI-augmented ways of working. They build cultures that encourage experimentation and learning rather than penalize those behaviors.

15. CHANGE MANAGEMENT IN THE GEN AI AGE

The Human Side of the Equation

Previous technology waves asked employees to learn tools. The Gen AI era demands more. It asks employees to collaborate with systems that learn, evolve, and in some respects behave unpredictably. The relationship is more dynamic, more ambiguous, and more consequential for how careers develop and how organizations grow.

Simply placing new technology in people's hands does not change how a company works. It creates the conditions for change, but only deliberate redesigning of how people adopt, use, and build confidence in AI will determine whether those conditions are realized.

The Gap Leaders May Not See

One of the more compelling findings in recent research is how different leaders and employees experience the AI moment. Leaders tend to underestimate how widely their teams already use Gen AI and overestimate how consistently it's used well. Employees are engaging with these tools at roughly three times the rate their leaders recognize.

When employees turn to consumer AI tools outside of enterprise platforms, company data can become exposed, and compliance risks grow quietly in the background, often without anyone intending harm. Without shared standards, results become uneven. Some employees unlock remarkable outcomes, while others produce work that, over time, can subtly erode the customer trust organizations have worked hard to build. Valuable insights stay siloed rather than benefiting the whole organization. Departments develop their own separate approaches; it becomes harder to build the kind of consistency that makes AI powerful.

The answer here isn't restriction. For most organizations, that window has already passed, and frankly, restriction alone was never the right response to a tool people find genuinely useful. The more productive path forward is structured: thoughtful training, clear governance, and well-defined pathways for responsible use that acknowledge where employees are, helping them go further with confidence.

Four Pillars of Lasting Adoption

IBM's research on AI adoption identifies four qualities that distinguish organizations where Gen AI takes hold from those where it stalls.

The first is trust: not trust in technology, but trust that AI is being deployed in ways that genuinely support employees rather than simply measuring or replacing them. When people believe the organization has their interests in mind, they engage. When they don't, they find workarounds.

The second is transparency: honest, ongoing communication about what AI is doing, how job functions are evolving, and how employees can raise questions or challenge recommendations. Transparency answers the question that every employee is quietly asking: What does this really mean for me?

The third is skills development, not just a one-time event. The capabilities required to work effectively alongside AI are dynamic, and the organizations that treat learning as a continuous system rather than a periodic initiative will develop workforces that are genuinely adaptive.

The fourth is agility: the willingness to deploy gradually, adjust based on what is learned, and lead with flexibility rather than rigidity. Technology failures rarely derail AI initiatives. The inability to learn and adapt does.

Cultural Reality

Perhaps the most instructive data point in recent AI research is this: most senior data leaders cite cultural challenges and change management as the primary obstacle to becoming AI-driven organizations. Technology is largely not the problem.

What that means in practice is that the hardest work of AI transformation is not architectural. It is behavioral. It is changing how people think, work, and relate to one another in an environment that is shifting around them.

Leaders who model AI openly, share their own learning journeys, acknowledge uncertainty, and demonstrate that adaptation is something they expect of themselves as well, create the conditions for others to follow. Organizations that build psychological safety around experimentation, that recognize employees who adopt thoughtfully and share generously, and that allow for the temporary productivity dip that genuine learning requires, tend to see something their impatient counterparts do not: lasting change.

Resistance to Gen AI is not monolithic and treating it as such is a common and costly mistake. It tends to be presented in three distinct forms, each requiring a different response.

Rational resistance emerges from legitimate concerns about AI accuracy, about bias, about privacy, and about job security. It deserves to be met with data, transparency, and evidence of concrete safeguards. Dismissing these concerns as technophobia misses the point and erodes the credibility leaders need to lead effectively.

Emotional resistance runs deeper. It is rooted in fear of obsolescence, attachment to familiar ways of working, and anxiety about an uncertain future. It cannot be argued away. It responds to empathy, to stories of peers who navigated the transition successfully, and to the quiet reassurance that comes from visible organizational support.

Political resistance tends to surface when AI shifts power dynamics, changes resource allocation, or alters organizational status. Engaging these stakeholders early, creating genuine opportunities for input, and aligning incentives with desired behaviors are more effective than any communication strategy alone.

In most organizations, all three forms of resistance exist simultaneously. The leaders who navigate the resistance well are those who diagnose before they prescribe.

Change Management as Competitive Advantage

The organizations that approach change management with genuine rigor allocating budget, leadership attention, and strategic urgency, alongside their technology investments consistently outperform those that treat it as an afterthought. The research is clear: they achieve higher revenue growth, greater cost savings, stronger employee satisfaction, and significantly higher rates of successful transformation.

This is not peripheral work. It is the work that determines whether Gen AI investments deliver on their promise or become another line item in a graveyard of well-funded initiatives that never scaled.

Closing

Leaders who apply the same rigor to change management that they apply to technology architecture and the same sense of urgency they reserve for competitive threats will move ahead, while others are left questioning why their transformation efforts are not delivering intended outcomes and leading to constant chaos and deteriorating employee engagement.

PART IV: THE SO WHAT

16. MEASURING WHAT MATTERS

"What gets measured gets managed" is one of the most frequently repeated phrases in business leadership. It endures because it is true. Measurement shapes behavior. When organizations measure the wrong indicators, they quietly steer teams toward the wrong outcomes.

For two decades, metrics like Average Handle Time (AHT), Cost per Contact, and First Contact Resolution (FCR) have been the dominant indicators of service performance. They served an important purpose in an era when contact centers were built to manage high volumes of transactional interactions at the lowest possible cost.

When efficiency metrics become the primary, or even the only, indicators of success, they begin to distort behavior across the organization. Associates shorten conversations instead of understanding the concern. Leaders approve investments that cut costs rather than improve experience. Vendors optimize service level agreements (SLAs) that look strong on paper but leave customers frustrated in practice. The numbers have improved. Customer trust quietly erodes.

Generative AI changes what we can see. For the first time, we can measure at scale what used to be invisible. Whether the AI answer was correct. Whether a customer's confidence held a week after the call ended. Whether the resolution promised to the customer materialized. The question is no longer whether these things can be measured. The question is whether we will measure them, and whether we will let what we find change how we lead.

The Limits of Efficiency Metrics

Traditional service metrics emerged in an era when operational visibility was limited and labor was the dominant cost. They assumed that efficiency, meaning reduced time and resources per interaction, was the primary path to value. Those assumptions no longer hold.

Consider the frontline associate who spends an extra four minutes comforting a customer dealing with a death in the family. Under AHT, she looks like a poor performer. From the customer's perspective, she just created a lifetime of loyalty.

Consider the chatbot that "deflected" 40 percent of calls. The deflection rate looks like a win but follow those customers and you may find many of them abandoning the issue, calling back days later, or closing their account.

Consider the vendor who consistently hits every SLA. On paper, they are performing but a detailed audit might suggest poor customer experience. In each case, the organization "wins" the metric and loses the relationship. This is the measurement paradox, and it is what the Gen AI era demands we fix.

Where Traditional Metrics Fall Short

The table below maps the metrics most CX organizations rely on today against the gaps they leave open, and the new metrics that close those gaps. The goal is not to replace the legacy measures. It is to see the full picture by placing quality indicators alongside efficiency indicators.

Traditional Metric	The Gap It Misses	Proposed New Metric
Average Handle Time (AHT)	Whether the customer felt their time was respected and the issue was truly addressed	**AI Customer Effort Score**
First Contact Resolution (FCR)	Whether the answer the customer received was correct	**Resolution Accuracy Rate**
Cost per Contact	The true cost of a quality outcome, not the cost of a completed interaction	**Experience Adjusted Cost per Resolution**
Post Interaction CSAT	Whether the customer's satisfaction survived contact with reality	**Confidence Durability Score**
AI Deflection Rate	Whether the deflected customer was genuinely served or simply turned away	**Regret Rate**
Standard QA Sampling	Whether AI responses are aligned with current policy and knowledge sources	**AI Accuracy Audit Score**
Employee Engagement Score	Whether frontline employees trust the specific AI tools in their workflow	**Employee Confidence in AI Tools**
Incident Response Time	Whether the issue could have been detected and resolved before the customer noticed	**Proactive Resolution Rate**

Every traditional metric was designed to answer a narrow operational question. None of them was designed to answer the question that matters most in a Gen AI environment. Did the customer receive an accurate, durable, and trustworthy outcome? The eight new metrics in the remainder of this chapter are the practical answer.

Eight Metrics for the Gen AI Era CX Scorecard

The eight metrics used in the Gen AI Era CX Scorecard are represented by tables that include what it measures, why it matters, how to calculate it, how to collect the data, and what scores represent a healthy range. These are not aspirational concepts. They are metrics a CX leader can stand up within one or two quarters using tools most organizations already have.

TABLE 1 - AI Customer Effort Score	
WHAT IT MEASURES	How easy it was for the customer to get their issue handled through AI channels.
WHY IT MATTERS	Customer Effort Score has been a reliable predictor of loyalty for more than a decade. Applied specifically to interactions handled by AI, it reveals whether your AI is genuinely helpful or simply pushing work back to the customer.
HOW TO CALCULATE	AI CES = Average response to the question: "How much effort did you personally have to put forth to handle your request today?" (1 = Very Low Effort, 7 = Very High Effort)
HOW TO MEASURE	a. Trigger a single question survey after any interaction handled primarily by AI. This includes chatbot, IVR, voice AI, and email response. b. Aim for a response rate of 10 percent or higher. Most enterprise survey tools support this natively. c. Segment by interaction type. Simple information request, transaction, complaint, technical support.
HEALTHY RANGE	Average score of 2.5 or lower. Scores of 4 or higher indicate the AI is creating friction rather than reducing it.

TABLE 2 - Resolution Accuracy Rate	
WHAT IT MEASURES	The percentage of cases marked "resolved" that were, in fact, resolved correctly.
WHY IT MATTERS	First Contact Resolution measures whether the customer called back. It says nothing about whether the answer was right. A customer who was told the wrong thing, accepted it, and never called back is counted as a success in FCR. They may have just given up on you.
HOW TO CALCULATE	Resolution Accuracy Rate = (Cases verified as correctly resolved ÷ Total cases marked resolved) × 100
HOW TO MEASURE	a. Pull a random sample of cases marked "resolved" each week. 100 to 300 cases is typically sufficient. b. For each case, verify downstream evidence. Did the refund post? Did the ticket service close without reopening? Did the billing dispute reverse correctly? Did the appointment happen? c. This is data you already have. It lives in the billing system, CRM, field service system, or core banking platform.
HEALTHY RANGE	92 percent or higher for standard transactions. 95 percent or higher for financial and regulated transactions.
WHAT TO WATCH	The gap between FCR and Resolution Accuracy Rate is a direct measure of how much resolution is being overstated. Pair this metric with Confidence Durability Score for the full picture.

TABLE 3 - Experience Adjusted Cost per Resolution	
WHAT IT MEASURES	The true cost of delivering a quality outcome, not just the cost of handling a contact.
WHY IT MATTERS	Cost per Contact treats every contact as equivalent. A correctly resolved case and a wrongly resolved case count the same, even though the wrongly resolved case will generate two to four downstream contacts, an escalation, or a churn event. This metric counts only the contacts that produced the outcome you wanted.
HOW TO CALCULATE	Experience Adjusted Cost per Resolution = Total service cost ÷ (Total contacts × Resolution Accuracy Rate × Confidence Durability Score)
HOW TO MEASURE	a. Start with your standard cost per contact calculation. b. Apply the Resolution Accuracy Rate and Confidence Durability Score as discount factors. A contact only counts as "productive" if it met the quality threshold. c. Report alongside standard Cost per Contact, not as a replacement. The gap between the two is the cost of low-quality service.
HEALTHY RANGE	This varies by industry, and the useful benchmark is internal. Compare your Experience Adjusted Cost per Resolution to your standard Cost per Contact. A ratio above 1.3 times means a significant share of your service spend is being wasted on interactions that did not produce real resolution.

TABLE 4 - Confidence Durability Score	
WHAT IT MEASURES	Whether a customer's confidence in the resolution they received holds up over time.
WHY IT MATTERS	A standard CSAT score captures the moment. It does not tell you whether the customer's confidence survived contact with reality. Whether the refund is posted. Whether the service call fixed the issue. Whether the policy change they were promised happened.
HOW TO CALCULATE	Confidence Durability Score = (Customers still confident at T+14 days ÷ Customers who reported confidence at T+0) × 100
HOW TO MEASURE	At the close of the interaction, ask one question. "How confident are you that your issue was fully resolved? (1 to 5)" Fourteen days later, send one question follow up to those same customers. "Looking back, was your issue resolved the way you expected? (Yes, Partially, No)" Score only customers who initially reported confidence, meaning a 4 or 5. Measure how many still say "Yes" at T+14.
HEALTHY RANGE	85 percent or higher indicates durable resolution. Below 75 percent indicates a false resolution problem. Customers are leaving interactions believing the issue is closed when it is not.
WHAT TO WATCH	The gap between initial CSAT and Confidence Durability is diagnostic. It tells you where your service system is selling promises it cannot keep.

TABLE 5 – Regret Rate	
WHAT IT MEASURES	The percentage of customers who, after an AI interaction, wish they had been connected to a human instead.
WHY IT MATTERS	This is the most honest question you can ask about your AI deployment. It cuts through efficiency theater and surfaces the gap between what the AI is doing and what the customer wanted.
HOW TO CALCULATE	Regret Rate = (Customers answering "Yes" to the regret question ÷ Total customers surveyed) × 100
HOW TO MEASURE	a. After an interaction handled by AI, ask. "Looking back, do you wish you had spoken with a person instead? (Yes, No)" b. Deploy monthly on a representative sample rather than continuously. Oversurveying customers degrades response quality. c. Cross-reference with interaction type, customer segment, and issue complexity.
HEALTHY RANGE	Below 15 percent overall. Below 8 percent for simple transactional interactions. Anything above 25 percent signals that AI is being deployed in the wrong contexts.
WHAT TO WATCH	Regret Rate tends to rise in inverse proportion to the emotional weight of the issue. Low regret on password resets paired with high regret on bereavement benefits is a sign your routing logic is uncalibrated.

TABLE 6 – AI Accuracy Audit Score	
WHAT IT MEASURES	The percentage of responses generated by AI aligns with current policy, product information, and approved knowledge sources.
WHY IT MATTERS	This is an important metric in a Gen AI CX environment and the one most organizations are not measuring. An AI system can be fast, friendly, and frequently wrong. Every traditional metric will tell you it is performing well. The failure described in Chapter 1 was an accuracy failure, and no efficiency metric catches it.
HOW TO CALCULATE	AI Accuracy Audit Score = (Audited AI responses aligned with current policy ÷ Total AI responses audited) × 100
HOW TO MEASURE	a. Pull a statistically representative weekly sample of interactions handled by AI. For most enterprise contact centers, 200 to 500 interactions per week is sufficient. b. Have QA analysts, subject matter experts, or a second AI model audit each response against the policy document, knowledge base version, or product specification in effect at the time of the interaction. c. Flag responses that are factually incorrect, that rely on outdated policy, or that are technically correct but misleading in context. d. Report weekly. Break the score out by intent category such as billing, claims, returns, and technical support so you can see where drift is happening.

HEALTHY RANGE	97 percent or higher for regulated industries such as financial services, healthcare, and insurance. 95 percent or higher for retail and general consumer business. Anything below 93 percent is an active risk.
WHAT TO WATCH	Accuracy degrades fast in the weeks following a policy change, product update, or pricing change. Build a trigger that forces a fresh audit whenever source material changes.

TABLE 7 – Employee Confidence in AI Tools	
WHAT IT MEASURES	Whether frontline employees believe the AI tools in their workflow are making them more effective.
WHY IT MATTERS	When employees do not trust the AI, they either override it, which eliminates the efficiency gain, or they defer to it even when they know better, which creates accuracy failures. Either way, you lose. Employee sentiment is the leading indicator of whether your AI deployment will deliver value.
HOW TO CALCULATE	Employee Confidence in AI Tools is the average score across three items on a quarterly pulse survey, each rated 1 to 5: *a.* "The AI tools I use give me accurate information." b. "The AI tools help me serve customers better than I could without them." *c.* "I would recommend our AI tools to a colleague starting in this role."
HOW TO MEASURE	Run as a quarterly pulse to all employees who serve customers. Keep the full survey to three or five questions. Response rates collapse beyond that. Segment by tenure and by specific AI tool, such as knowledge assistant, call summarization, and response drafting.
HEALTHY RANGE	Average of 4.0 or higher. A score below 3.5 is a warning that the tools are creating drag rather than leverage.

TABLE 8 – Proactive Resolution Rate	
WHAT IT MEASURES	The percentage of issues the organization detects and resolves before the customer ever contacts service.
WHY IT MATTERS	The most valuable service interaction is the one the customer never had to initiate. Modern telemetry, billing analytics, and account monitoring make this achievable, but only when it is measured as a deliberate outcome
HOW TO CALCULATE	Proactive Resolution Rate = (Issues detected and resolved before customer contact ÷ Total issues identified) × 100
HOW TO MEASURE	Tag each resolved issue by origin. Was it customer initiated or system initiated? Track proactive interventions. Outbound notifications, automatic credits, preemptive service dispatch, and account self-healing actions all count. Credit an intervention as "proactive resolution" only if it was initiated before the customer contacted you, and only if it prevented a contact that would otherwise have occurred. This can be validated through contact suppression analytics.
HEALTHY RANGE	15 percent or higher for mature programs. Most organizations start below 5 percent. Moving from 5 percent to 20 percent typically produces a 10 to 15 percent reduction in overall contact volume

Rethinking Talk Time

One global bank illustrates how measurement changes behavior. The organization replaced traditional AHT targets with a blended metric called Time Well Spent. Rather than measuring call duration alone, the new metric evaluated whether customers felt their time was respected. Employees were assessed on both resolution quality and the emotional tone of the conversation, with AI enabled dashboards tracking sentiment alongside efficiency.

The results were significant. Service costs declined as repeat contacts dropped. Customer loyalty scores improved. Employees reported reduced stress because they were no longer forced to truncate interactions that required empathy and attention. The lesson is simple. When you measure the right outcomes, employees adopt the behaviors that produce them. When you measure the wrong ones, they work around you.

What the Chapter 1 failure Teaches us About Measurement

Return to the failure described in Chapter 1. Ask which of the eight metrics above would have caught it.

AHT would not have. The interaction was efficient. First Contact Resolution would not have. The case was marked closed. CSAT at the end of the call would have scored well. The AI deflection rate would have counted it as a successful self-service event. Every traditional dashboard would have declared the system healthy.

But the AI Accuracy Audit Score would have caught it. The AI was citing outdated policy, and a weekly audit against the current policy document would have surfaced the gap. The Confidence

Durability Score would have caught it. The customer's confidence collapsed within two weeks when reality contradicted the promise. The Resolution Accuracy Rate would have caught it. Downstream evidence showed the case was not actually resolved.

The most dangerous failures in a Gen AI service environment are not the ones that show up as complaints. They are the ones that hide inside good-looking numbers. The Trust Scorecard is designed to make those failures visible before they compound.

Why Measurement Matters at the Leadership Level

For CX leaders, measurement is not simply an operational tool. It is the instrument that determines whether your work receives sustained investment. Boards and executive teams increasingly demand clear connections between customer experience performance and financial outcomes.

Each of the eight metrics above maps to a specific business outcome your CFO cares about.

- AI Customer Effort Score
- Resolution Accuracy Rate
- Experience Adjusted Cost Per Resolution
- Confidence Durability Score
- Regret Rate
- AI Accuracy Audit Score
- Employee Confidence in AI Tools
- Proactive Resolution Rate

The Leadership Mandate

Transforming your measurement system is not a technology project. It is a leadership decision. The work is practical.

Redesign the dashboard. Place trust, confidence, and accuracy metrics alongside efficiency metrics, not beneath them. Retire vanity metrics that do not reflect customer relationships.

Educate the board and the frontline. Boards need to understand what these metrics protect them from. Frontline teams need to understand how these metrics protect them from being measured on the wrong things.

Hold vendors accountable for outcomes, not just for SLAs. A vendor hitting every service level while degrading your AI Accuracy Audit Score is not partnering with you. They are compounding your risk.

Recognize the behaviors the new metrics produce. Celebrate the associate who preserved the relationship. Celebrate the AI engineer who caught the accuracy drift. What leaders reward, teams repeat.

The organizations that will lead customer experience in the Gen AI era are not the ones with the fastest systems or the lowest cost per contact. They are the ones willing to measure what matters. Accuracy. Durability. Confidence. Trust. They are the ones willing to make those numbers the scoreboard the whole organization plays on.

THE TRUST SCORECARD

The Trust Scorecard Diagnostic: Where Do You Stand?

Assess your organization's maturity measurement by answering the scorecard diagnostic questions. For each statement, rate your organization on a scale of 1 (Not at all) to 5 (Fully in place). Total your score within each pillar and across all four to identify your strengths and gaps.

THE TRUST SCORECARD Organizational Self-Assessment

Dimension	1 Critical Risk	2 Developing	3 Capable	4 Strong	5 Exemplary
Trust Index *Perceived fairness, transparency & accuracy*	Customers feel misled or treated unfairly	Trust is fragile; fairness is inconsistent	Most interactions perceived as fair and honest	Customers feel consistently heard and respected	Deep trust; customers actively advocate for the brand
Customer Confidence Score *Confidence that issues were resolved correctly*	Customers leave uncertain; high re-contact rate	Confidence varies; partial resolution is common	Most customers confident in outcome accuracy	High confidence; low repeat-contact rate	Near-total confidence; follow-up is standard practice
Return on Experience (CX ROI) *CX improvements tied to measurable financial outcomes*	No link between CX and financial outcomes	Anecdotal link exists; no formal measurement	Basic linkage tracked; limited exec visibility	CX metrics tied to retention and revenue growth	CX ROI drives strategic investment decisions
Employee Empowerment Index *Whether AI tools enhance (vs. threaten) frontline staff*	Staff feel threatened or replaced by AI	Inconsistent adoption; low tool confidence	Most employees use AI tools adequately	Employees feel augmented and supported by AI	AI actively enhances performance and satisfaction
Proactive Resolution Rate *Issues resolved before customers report them*	All issues reactive; customers always initiate	Occasional proactive outreach; no systematic process	Some predictive capability applied inconsistently	Regular proactive resolution before customer contact	Anticipatory service; issues prevented systematically
Knowledge Accuracy & AI Transparency *AI response accuracy and explainability to customers*	AI responses frequently inaccurate or outdated	Accuracy varies; no audit process in place	Regular audits; moderate accuracy maintained	High accuracy; AI reasoning clear to customers	Near-perfect accuracy; full transparency is standard

1 = Critical Risk 2 = Developing 3 = Capable 4 = Strong 5 = Exemplary

Use this scorecard to benchmark your organization's current trust posture and identify priority dimensions for Gen AI investment.

The Trust Scorecard: What to Measure and How

Throughout this book, we have explored how generative AI reshapes the customer experience from proactive resolution and empathy-driven interactions to agentic workflows and governance architectures.

Leaders benefit from a measurement framework that links operational metrics to the outcomes that matter most to boards, CFOs, and customers.

This is the purpose of the Trust Scorecard.

The Trust Scorecard is not another dashboard to monitor. It is a decision-making framework that answers a deceptively simple question: Is our AI investment making customers trust us more, and can we prove it?

Implementing the Trust Scorecard

Most organizations already collect the raw data for at least half of the metrics described above. The work is not in inventing new measurements; it is in connecting existing data into a coherent narrative. Begin by mapping current metrics. Identify which pillar has the weakest coverage; that is where to focus first.

Establish a cadence: The Trust Scorecard should be reviewed monthly at the operational level and quarterly at the executive and board level. Use operational reviews to identify trends and executive reviews to connect CX performance to strategic priorities and investment decisions. Set targets that reflect trust, not just efficiency. Rather than targeting a ten percent reduction in handle time, aim for a ten-point improvement in Customer Confidence Score while maintaining or improving cost efficiency. This subtle shift in framing changes the incentive structure.

Interpreting Your Score

64–80: Trust Leader. Your measurement architecture is mature and integrated. Focus on predictive analytics and industry benchmarking. You are positioned to use trust as a competitive differentiator.

48–63: Trust Builder. You have strong foundations in some pillars but gaps in others. Prioritize the lowest-scoring pillar and build the connections between metrics. The framework is there; the integration work remains.

32–47: Trust Aware. You recognize the importance of trust-based measurement but are still relying heavily on traditional efficiency metrics. Start with the three easiest wins in your lowest pillar and build momentum.

Below 32: Trust Opportunity. Your measurement system is optimized for a pre-AI era. This is not a criticism — it is an opportunity. The organizations that close this gap fastest will have the greatest competitive advantage. Start with Pillar 1 and build outward.

The Trust Scorecard is designed to evolve with your organization. As your AI capabilities mature, the metrics within each pillar will shift from lagging indicators to leading ones, from descriptive to predictive. The architecture stays the same. The ambition grows.

Closing

Generative AI unlocks more than efficiency; it enables a deeper understanding of what customers truly value: trust, confidence, and genuine care. Organizations that embrace this shift discover that meaningful outcomes extend beyond cost savings or handle time. They appear in stronger relationships, deeper loyalty, and the confidence customers carry long after an interaction ends.

17. LEADERSHIP IN THE AGE OF INTELLIGENT EXPERIENCE

Technology does not transform organizations. Leaders do. Every major inflection in business history: industrialization, globalization, and digitization were determined not by the tools available but by the leaders willing to rethink operating models, redefine value, and move people through uncertainty with clarity and conviction. Generative AI is no different. Its promise is substantial, but its impact will be decided by leadership choices: what gets prioritized, what gets protected, what gets scaled, and what gets embedded into culture as "the way we work."

For decades, many executives were rewarded primarily for managing efficiency. COOs optimized processes and throughput. CCOs reduced complaints and improved service levels. Transformation leaders drove reductions in cost-to-serve and productivity improvements. Those capabilities still matter, but they are no longer sufficient. In an era where customers can switch brands with a click and reputations can be damaged at the speed of a viral post, efficiency without trust is a fragile strategy.

The Gen AI era expanded the mandate: leaders are now responsible for architecting experiences that deepen loyalty, reduce risk, and generate sustainable growth. The question is no longer simply, "How fast did we resolve the issue?" It is, "Did we strengthen trust while resolving it, and can we prove it?" High-performing leaders stop counting transactions as the definition of value and start measuring relationship outcomes: confidence, clarity, fairness, and emotional resonance. Efficiency becomes the baseline; trust becomes the differentiator.

This shift demands a new set of leadership competencies, muscles that many organizations have not historically developed at scale. AI literacy without jargon. Leaders do not need to be data scientists, but they must understand how generative systems behave, where they fail, and what responsible deployment requires. That includes a working understanding of model limitations, data dependencies, hallucination risk, bias exposure, and the operational controls that prevent intelligent errors from becoming customer harm. Most importantly, leaders must serve as translators able to connect technical realities to frontline workflows and human implications in a way that builds confidence rather than confusion. They do so by using competencies of empathy at scale, risk courage, and storytelling with data.

Empathy at scale: That means designing journeys where customers feel understood rather than processed and designing workflows where employees feel supported rather than monitored. It also means resisting the temptation to optimize purely for containment and deflection at the expense of the relationship.

Risk courage: the ability to move forward responsibly without letting caution become paralysis. Regulated industries are right to treat trust as an asset that cannot be compromised. But the leaders who create advantage are those who build clear boundaries, governance mechanisms, and auditability. They should then execute with intention inside those guardrails. Risk courage is not recklessness; it is the discipline to design controls early, test aggressively, and scale only when reliability and accountability are demonstrable.

Storytelling with data: Gen AI transformations produce

dashboards, metrics, and pilots, but dashboards do not mobilize organizations. Leaders must connect AI initiatives to outcomes that resonate with boards, employees, and customers: growth, retention, cost-to-serve efficiency, risk reduction, and brand strength. This requires narrative skill: an ability to translate metrics into meaning, to connect experience changes to financial outcomes, and to tell a coherent story of how AI strengthens the enterprise rather than simply modernizing it. In the Gen AI era, influence belongs to leaders who can prove value and inspire belief simultaneously.

One of the most effective AI leaders I have worked with made a simple decision early in her Gen AI deployment: she added a single line to every team member's performance review. It read: "*contributes to the accuracy and currency of team knowledge systems*".

It was not weighted heavily. It was not tied to compensation, but it changed behavior overnight. Associates who had previously treated the knowledge base as someone else's responsibility began flagging outdated articles. Team leads started reviewing the knowledge management portal weekly. Quality analysts, who had always measured associate behavior, began measuring the accuracy of the information.

Within six months, knowledge accuracy, measured by the percentage of AI-suggested responses that matched verified current policy, improved from seventy-one percent to ninety-four percent. Repeat contact rates and escalations decreased. The team developed a sense of collective ownership over the system that served their customers.

She told me the idea came from the Chapter 1 failure pattern: the recognition that when nobody owns knowledge, everybody suffers. "I could not control the model," she said. "But I could control the culture around the information the model was using. That turned out to be the thing that mattered most." Change fails when it is sold as certainty or imposed as a mandate. It succeeds when leaders acknowledge what is shifting, explain why it matters, and create a credible path for people to grow with and a future worth building together. When leaders treat adoption like a product, they design it, measure it, refine it, and invest in it with the same rigor applied to the technology itself. This is where the "human multiplier" becomes central.

One of the most common strategic mistakes is treating AI as a substitute for leadership. The opposite is true: as AI becomes more capable, the role of leadership becomes more impactful. AI can draft responses, but leaders determine what "good" looks like and where human judgment must remain. AI can optimize workflows, but leaders decide whether those workflows reflect the organization's values, fairness, transparency, and care—or whether they quietly trade trust for speed. AI can measure sentiment and predict churn, but leaders must interpret the signals, act, and close the loop across product, policy, and operations. The more technology expands what is possible, the more leadership determines what is acceptable and what is wise.

Accountability is already shifting accordingly. Boards, regulators, and stakeholders are asking new questions that extend beyond quarterly performance: How are you using AI responsibly? How are you protecting customer data and privacy? How are you preventing bias and ensuring fairness? How are you preparing your workforce for AI-augmented work? These are not peripheral questions. They are core indicators of organizational maturity and trustworthiness. In the coming years, executives will be evaluated not only on revenue growth and cost performance, but on their ability to steward trust through governance, transparency, and human-centered design.

That is why legacy in the Gen AI era will look different. Some leaders will be remembered for automation programs that reduced headcount, shortened handle times, and cut costs—wins that may look impressive in quarterly reports but do not necessarily endure. Others will be remembered for building organizations where customers felt genuinely seen, employees felt meaningfully empowered, and shareholders benefited from resilient growth rooted in trust. The leaders who approach generative AI not merely as a tool for efficiency, but as a trust amplifier and experience differentiator, will build capabilities that competitors cannot easily replicate—because culture, governance, and credibility are far harder to copy than technology.

The age of intelligent experience is not about machines outsmarting humans. It is about leaders re-imagining what becomes possible when humans and machines work together with clarity, integrity, and purpose. The defining leadership capability will be the ability to blend vision with empathy, operational discipline with innovation, and data-driven performance with human-centered trust. The future of customer experience will not be determined by the most advanced algorithm. It will be defined by the leaders with the courage to build responsibly, scale thoughtfully, and put humanity at the center of intelligent systems.

18. THE NEXT FRONTIER: AI + HUMANS CREATING VALUE

For most of modern business history, the relationship between companies and their customers has been fundamentally transactional. Organizations designed products, customers purchased them, and service functions existed primarily to resolve problems that arose after the purchase. Value flowed in one direction: from company to consumer. Even as digital transformation improved speed and convenience, the underlying dynamic remained largely unchanged. Companies still created the experience; customers simply received it.

The emergence of generative AI is beginning to reshape that model in a profound way. Customers no longer occupy the role of passive recipients of products and services. Increasingly, they are becoming active participants in shaping the experiences they consume. This shift represents the next frontier of customer experience: a move from service delivery to value co-creation, where organizations and customers collaborate continuously to co-design outcomes. Generative AI serves as the enabling infrastructure that makes this level of collaboration possible at scale.

From Reactive Service to Collaborative Experience

Customer service has evolved through distinct phases over the past decades. In its earliest form, service was mainly reactive. Customers encountered problems, contacted support, and associates resolved the issue as quickly as possible. The goal was operational closure; fix the problem and move on.

As digital channels matured and data analytics improved, organizations shifted toward a proactive service model. Companies learned to anticipate issues, notify customers before disruptions, and intervene earlier in the relationship lifecycle. Proactive service reduced friction and improved satisfaction, but it still maintained the traditional dynamic in which companies designed the experience, and customers responded to it.

The next phase goes further. Generative AI enables organizations to adopt a collaborative model in which customers participate directly in designing their own experiences. Instead of predicting what customers want based solely on historical data, companies can invite customers to define their needs dynamically, with AI translating those inputs into personalized outcomes in real time.

In this environment, personalization evolves into co-design. Rather than organizations relying solely on algorithms to infer preferences, customers can actively guide the experience through interactive AI systems. Feedback loops also become richer. Instead of static surveys that are reviewed weeks later, AI systems can incorporate customer feedback immediately, learning from each interaction and adapting future experiences accordingly.

Innovation itself begins to change as well. When customers can simulate outcomes, experiment with options, and explore alternatives through AI-enabled platforms, they effectively become participants in the innovation process. Ideas that once required months of research and development can emerge organically through millions of real-time interactions between customers and intelligent systems.

Co-Creation in Practice

The potential of co-creation is already visible across multiple industries, though generative AI is dramatically expanding its scope.

In banking and financial services, for example, customers increasingly interact with AI-powered financial planning tools that allow them to explore life scenarios dynamically. Instead of receiving static financial advice, individuals can test different paths like saving more aggressively, purchasing additional property, adjusting investment allocations and immediately seeing the projected outcomes. In this model, the financial advisor's role evolves from information provider to strategic guide, helping customers interpret possibilities rather than simply delivering recommendations.

Retail offers another compelling example. AI-driven shopping environments allow customers to build highly personalized product combinations from clothing collections to curated meal plans based on lifestyle preferences, health goals, and budget considerations. Retailers now do more than simply present product catalogs: they enable customers to assemble experiences that reflect their individual identities. Over time, the retailer becomes less a seller of goods and more a partner in shaping how customers live, eat, dress, and travel.

Healthcare also illustrates the transformative potential of collaborative engagement. Patients can use AI tools to organize symptoms, review potential diagnoses, and generate informed questions before visiting a physician. By the time the appointment begins, both patient and doctor enter the conversation with a shared understanding of the situation. Physicians spend less time gathering basic information and more time exercising clinical judgment and empathy. The relationship becomes a collaborative diagnostic process rather than a one-sided consultation.

Even traditional customer support interactions are evolving toward co-creation. When issues are resolved, AI systems capture customer insights into which aspects of the process were helpful, confusing, or unnecessary. These insights feed directly into continuous improvement systems that refine service workflows in near real time. Over time, service journeys become adaptive systems shaped by the cumulative experiences of millions of customers.

Why Co-Creation Strengthens Loyalty

The shift toward collaborative engagement has implications that extend far beyond convenience or personalization. Co-creation fundamentally alters the psychological relationship between companies and customers.

When customers participate in shaping their own experiences, they gain a sense of agency that traditional service models rarely provide. They feel empowered because they no longer simply receive service; they shape the outcomes. They feel respected because their perspectives visibly shape the organization's decisions. They feel connected because the relationship begins to resemble a partnership rather than a transaction. These emotional dynamics create loyalty that is significantly stronger than that generated by promotional incentives or reward programs. Discounts can attract customers temporarily, but a partnership builds long-term commitment. Customers who feel invested in the experiences they help shape are far less likely to leave when competitors offer marginally better pricing or convenience.

Leadership Implications

For COOs, CCOs, and transformation leaders, enabling co-creation requires a fundamental shift in organizational thinking. Traditional service organizations are built around processes designed for efficiency and control. Co-creation requires designing platforms that allow participation and adaptability.

Leaders must begin thinking less about fixed processes and more about experience platform systems that enable customers to dynamically influence outcomes while maintaining operational reliability and compliance. Transparency also becomes critical. When organizations invite customer input, they demonstrate how that input influences decisions. Nothing erodes trust faster than feedback that appears to disappear into a void.

Another leadership challenge lies in balancing automation with customer agency. AI systems can guide experiences efficiently, but customers must retain meaningful control over decisions that affect them. If automation becomes overly prescriptive, the sense of partnership disappears. Leaders must therefore design experiences where AI assists rather than dictates.

Equally important is protecting customers from complexity. Co-creation should not feel like work. The best systems allow customers to shape outcomes intuitively, without requiring them to navigate complicated interfaces or decision trees. Simplicity remains essential even as participation increases.

Lessons from Early Co-Creation Models

Elements of co-creation have existed for years in certain business models. Online marketplaces, for example, have long used customer reviews and recommendation systems to shape future purchasing journeys. These mechanisms give customers influence over how others discover and evaluate products. Technology ecosystems provide even stronger examples. Platforms that enable external developers to build applications extend innovation beyond a single company. In these ecosystems, the organization provides the infrastructure while the community contributes ideas, capabilities, and creativity.

Generative AI dramatically expands these models. Instead of limiting co-creation to product reviews or developer communities, AI enables collaborative engagement in every interaction; purchasing decisions, service experiences, product development feedback, and long-term relationship management.

Customers as Strategic Partners

The implications of this shift are significant. The traditional business model positioned companies as creators and customers as consumers. In the emerging model, collaboration exists between organizations and the people they serve, producing value. Strategy, therefore, evolves in stages. It moves from a product-first mindset, where companies focus primarily on what they build, to an experience-first mindset that prioritizes how customers interact with those products. It advances towards a partnership-first mindset in which organizations and customers work together to continuously generate value.

In this environment, the sources of innovation expand dramatically. Organizations no longer face limits from ideas generated solely within their internal teams. Instead, they benefit from insights emerging across an entire ecosystem of engaged customers whose experiences shape future offerings.

Closing

The next frontier of customer experience emerges not from companies doing more for customers, but from companies doing more with them. Generative AI makes this collaboration possible at a scale that was previously unimaginable. By turning interactions into dialogue and personalization into partnerships, organizations can transform service relationships into ecosystems of shared value.

Companies that embrace this evolution will discover that their competitive advantage no longer comes solely from the products they create or the services they deliver. It comes from the communities they cultivate—communities built on trust, collaboration, and continuous innovation.

In the future, successful organizations will realize they have more than traditional customers. They will have co-creators.

19. INDUSTRY-SPECIFIC TRANSFORMATION INSIGHTS

The generative AI revolution is unfolding unevenly across industries. While certain principles such as the need for trustworthy data, thoughtful governance, and human-centered design apply universally, each sector faces its own regulatory pressures, operational realities, and customer expectations. As a result, the path to AI-enabled transformation varies by industry context. Financial services, retail, healthcare, manufacturing, and telecommunications are each approaching this shift through the lens of their own competitive pressures and structural constraints. For leaders, understanding these sector-specific dynamics is essential, as the organizations that move first in an industry often set the standards that competitors later follow.

Financial Services: Trust Operating at Digital Speed

Financial institutions have been among the earliest adopters of advanced data analytics and artificial intelligence. More than seventy-five percent of banks now plan to increase investments in cloud infrastructure and data management platforms to support generative AI initiatives. Despite this maturity, financial service organizations face one of the most challenging environments for innovation due to strict regulatory oversight and the fundamental importance of trust in financial relationships.

Successful implementations in this sector typically begin with compliance-first architecture. Financial institutions must design AI systems with transparency, auditability, and explainability built directly into their operating frameworks. Regulators increasingly expect organizations to demonstrate how algorithmic decisions occur and how risks are mitigated. Attempting to retrofit compliance after deployment often proves more difficult than embedding governance mechanisms during initial development.

Equally important is balancing personalization with privacy. Customers increasingly expect financial advice tailored to their specific circumstances, yet they remain cautious about how personal data supports those recommendations. Organizations that clearly explain which data informs guidance—and give customers control over how their information supports those insights—often build deeper trust and stronger engagement.

Fraud detection represents another area where AI provides immediate value. By analyzing transaction patterns across enormous datasets, AI systems can identify suspicious behavior in real time while reducing the number of false positives that inconvenience legitimate customers. The most effective financial institutions use AI to manage routine monitoring tasks while preserving human expertise for complex advisory relationships. When digital channels handle routine transactions, human advisors gain the time and context needed to support customers during major life events such as retirement planning, home purchases, or wealth management decisions.

One large financial institution recently implemented generative AI tools that allow customers to explore different financial planning scenarios interactively. Individuals can test how adjusting monthly savings contributions affects retirement outcomes or how purchasing additional property might influence long-term financial stability. Advisors remain available for strategic guidance, while customers take on a more active role in understanding their financial future. The result was a significant increase in engagement with financial planning tools and improved satisfaction with advisory services.

Retail and Consumer Businesses: Experience as Competitive Differentiation

Retail and consumer brands face a different challenge. Competition is intense, switching costs are low, and customer expectations for personalization continue to rise. Surveys consistently show that more than seventy percent of consumers want companies to understand their preferences and tailor experiences accordingly. Yet only about forty percent of customers believe the experiences they receive are genuinely personalized. This gap between expectation and perception represents both a risk and a strategic opportunity.

Retail organizations that succeed with AI focus on removing friction from the customer journey. Returns management, for example, has historically been one of the most frustrating aspects of retail transactions. Generative AI enables retailers to streamline the entire process, allowing customers to initiate returns through mobile devices, verify eligibility instantly, and receive refunds within minutes. When returns become effortless, what was once a loyalty-damaging experience becomes a differentiating advantage.

AI is also transforming inventory management. Retailers can analyze demand patterns at extremely granular levels, predicting not only which products will sell but also the specific sizes, colors, and locations where demand will occur. This capability reduces costly stockouts and excess inventory while improving working capital efficiency.

Conversational commerce represents another emerging shift. Instead of navigating extensive product catalogs, customers increasingly interact with AI systems that interpret natural-language requests and present curated recommendations. A shopper might simply describe a need such as hiking boots suitable for wide feet under a specific budget and receive highly relevant options instantly. These systems reduce decision fatigue while improving conversion rates. One fashion retailer introduced AI-powered virtual stylists that can recommend complete outfits based on a customer's lifestyle, preferences, and budget. The system explains why particular items work together and adapts recommendations based on feedback. In this scenario, customers purchased items better suited to their needs, and return rates declined significantly while customer satisfaction improved.

Healthcare: Precision Care Balanced with Human Judgment

Healthcare organizations face perhaps the most complex environment for adopting generative AI. The stakes involve not only financial outcomes but human health and safety. Strict privacy regulations, clinical accountability, and deeply personal patient interactions all influence how AI can be deployed responsibly. For many healthcare systems, the most immediate opportunity lies in administrative automation. Tasks such as insurance verification, appointment scheduling, prescription refills, and billing inquiries consume significant staff time that could otherwise be devoted to patient care. AI systems can efficiently manage these routine processes, allowing clinicians to focus on diagnosis, treatment planning, and patient engagement.

Generative AI also supports clinical decision-making by summarizing patient histories, identifying potential drug interactions, and surfacing relevant research. Importantly, these tools function as decision-support systems rather than replacements for physician expertise. Medical professionals retain ultimate responsibility for clinical judgments.

Patient engagement has become another important area of innovation. AI assistants can help patients prepare for appointments by documenting symptoms, organizing questions, and providing guidance on following treatment plans after visits. This preparation improves both the quality of consultations and the likelihood that patients adhere to recommended care.

One health system introduced an AI-based pre-visit preparation tool that allows patients to describe symptoms in natural language before meeting with a physician. The system generates structured summaries for clinicians, enabling them to quickly review patient information. Physicians reported spending less time gathering background details and more time discussing diagnosis and treatment options with patients.

Manufacturing: Intelligence in Operational Systems

Manufacturing organizations have historically focused AI adoption on operational efficiency rather than customer-facing experiences. Roughly eighty percent of manufacturers now use AI tools in some capacity to optimize supply chains and production processes. However, many still struggle to connect strategic AI ambitions with day-to-day operational realities.

Predictive maintenance represents one of the most widely adopted applications. Sensors embedded in industrial equipment generate vast streams of operational data that AI systems analyze to identify early signs of mechanical failure. Maintenance can be scheduled during planned downtime rather than after costly breakdowns.

Supply chain resilience has also become a major focus. AI models can simulate thousands of scenarios involving supplier disruptions, transportation delays, or demand fluctuations. These insights allow companies to adjust sourcing strategies and inventory positioning before disruptions occur.

Quality control has benefited from advances in computer vision technology. AI-powered inspection systems can detect defects in products far more consistently than manual inspections, improving product quality while reducing waste.

Even customer service functions in manufacturing are evolving. For companies that produce complex industrial equipment, AI tools assist service teams in diagnosing technical problems remotely, recommending solutions, and ensuring technicians arrive on-site with the correct parts and expertise.

Telecommunications: Intelligent Networks and Proactive Support

Telecommunications providers operate complex infrastructures in the modern economy. They must manage vast networks, serve millions of customers simultaneously, and maintain consistent service quality across diverse geographic regions.

AI is transforming how telecom companies monitor and optimize network performance. Advanced analytics systems analyze traffic patterns in real time and dynamically adjust bandwidth allocation, routing decisions, and capacity management. These capabilities help maintain reliable service during periods of heavy demand.

When outages occur, AI-driven systems can identify affected customers immediately and provide proactive updates regarding service restoration timelines. This targeted communication significantly reduces customer frustration and lowers inbound call volumes during disruptions.

Generative AI also enables more personalized service plans. By analyzing usage patterns across voice, data, and digital services, AI systems can recommend plans that better match customer behavior. In some cases, these recommendations even lower customer costs, demonstrating transparency and strengthening loyalty.

Another important application involves churn prevention. AI systems can identify subtle signals indicating that a customer may be considering switching providers—such as changes in usage patterns, billing disputes, or repeated service issues. Telecom companies can then intervene proactively with retention offers or service improvements before the customer leaves.

Lessons Across Industries

Despite the unique characteristics of each sector, patterns emerge across successful AI transformations. Organizations that generate meaningful value typically begin with specific customer or operational pain points. Clear problem definitions guide more focused investments and measurable outcomes.

Equally important is preserving human escalation pathways. Regardless of industry, certain situations require empathy, judgment, or contextual understanding that AI cannot fully replicate. Ensuring customers can easily transition from automated systems to human assistance remains essential.

Trust must also remain a central metric. Efficiency improvements are valuable, but they lose meaning if customers feel alienated or unsupported. Leading organizations track both operational performance and trust indicators simultaneously to ensure that efficiency gains do not undermine relationships.

Finally, employee readiness determines whether AI initiatives succeed or fail. Technology adoption depends on whether employees understand, trust, and effectively use new tools. Training, communication, and cultural alignment, therefore, play as significant a role in transformation as technology itself.

Closing

The organizations that master AI-enabled customer experience within their industries will not simply gain temporary advantages. They will redefine customer expectations. Once customers experience proactive service, personalized engagement, and intelligent support from a single company, they begin to expect the same from every competitor.

For leaders, the strategic question is clear. Will your organization set the standard for AI-driven customer experience in your industry, or will it spend the next several years trying to catch up to those who moved first?

20. THE PATH FORWARD

If 2025 represented a moment of recalibration for artificial intelligence when the initial excitement surrounding generative AI collided with operational realities, then 2026 will be remembered as the year the technology became practical. The conversation is moving away from the pursuit of ever-larger models and toward the far more demanding task of making AI systems usable, trustworthy, and deeply embedded within everyday business operations. The real work of AI transformation is no longer about proving that technology is impressive. It is about proving its value.

From Hype to Pragmatic Execution

The early experimental phase of generative AI is coming to an end. In the past years, many organizations have launched pilot programs and innovation labs designed to explore what technology might eventually enable. Those initiatives were valuable for building awareness and early capability, but they also led to a proliferation of disconnected experiments with limited operational impact. As budgets tighten and expectations rise, executives are increasingly demanding measurable results. Projects that cannot demonstrate clear business value face reconsideration, while those tied to tangible outcomes continue to receive sustained investment.

This shift toward pragmatic execution appears across multiple dimensions. One of the most notable is the growing recognition that the greatest early value of AI often lies behind the scenes rather than in high-profile customer-facing applications. Internal operations such as knowledge management, employee productivity, workflow automation, and decision support offer opportunities where AI can deliver measurable improvements quickly. These operational gains may not generate headlines, but they often have the greatest financial and organizational impact.

Another important change is the growing discipline around problem definition. Early initiatives began with the question, "Where can we apply AI?" Today, more successful organizations start with a different question: "What specific problem are we trying to solve? This subtle shift in framing dramatically improves the likelihood of success.

Equally critical is the recognition that AI adoption is fundamentally a human transformation rather than a purely technical one. Organizations that invest heavily in technology but neglect workforce preparation frequently find their tools underutilized or misunderstood. By contrast, companies that pair technological investments with thoughtful change management, training programs, and cultural alignment see far stronger adoption rates and performance improvements.

The Three Factors That Separate Success from Stalled Pilots

Research analyzing enterprise AI adoption through 2025 consistently identifies three factors that distinguish organizations achieving real value from those stuck in pilot mode. Interestingly, these factors have less to do with algorithm sophistication than with organizational readiness and leadership discipline.

The first factor is **Organizational Readiness**. Even the most advanced AI system cannot deliver meaningful value if the organization deploying it lacks the infrastructure, processes, and cultural alignment necessary to support it. Successful organizations invest in foundational capabilities before pursuing advanced applications. This includes modern data architectures that make information accessible and reliable, operational processes designed to incorporate AI insights rather than resist them, and employees who understand both the capabilities and limitations of AI tools.

Companies have launched numerous AI pilots simultaneously to explore possibilities across multiple departments. While this approach generated ideas, it often diluted resources and attention. Organizations that succeed typically concentrate their efforts on a limited number of high-impact domains: customer on-boarding, supply chain optimization, or workforce productivity.

Cultural Readiness is equally important. Organizations that reward experimentation, encourage cross-functional collaboration, and treat early failures as learning opportunities tend to scale successful AI initiatives faster. Metrics also evolve to support this readiness. Instead of focusing solely on efficiency improvements, leading organizations measure outcomes such as trust, adoption, and long-term customer value.

A disciplined focus allows teams to redesign workflows, integrate data sources, and build governance structures that support sustained deployment. A single, thoroughly executed transformation often delivers more value than multiple pilots that never progress beyond experimentation.

The third factor is the creation of **Trust Infrastructure** alongside technical infrastructure. Trust requires consideration from the very beginning of AI deployment. Customers, employees, regulators, and investors increasingly expect organizations to demonstrate that AI systems operate transparently and responsibly. Companies that embed governance mechanisms from the beginning, such as clear explanations of AI decisions, accessible human escalation pathways, robust privacy protections, and regular bias testing consistently outperform those that attempt to add safeguards later.

Trust infrastructure also includes giving customers meaningful control over their interactions with AI systems. When individuals understand when AI is being used and retain the ability to request human assistance, they are far more likely to accept and even prefer AI-enabled services.

Workforce Transformation

Another defining feature of the next phase of AI adoption involves the transformation of work itself. Research from the World Economic Forum projects the emergence of approximately 170 million new roles globally between 2025 and 2030, alongside the displacement of about ninety-two million existing roles. This shift produces not a contraction of the workforce, but a net expansion of seventy-eight million positions, with a significant share focused on designing, guiding, and governing intelligent systems.

For leaders, this transformation creates a profound reskilling imperative. Rather than relying solely on hiring new technical talent, organizations must invest in helping existing employees adapt to new responsibilities. Workers need clear career pathways that demonstrate how current roles can evolve alongside AI capabilities. Companies that provide transparent development opportunities and meaningful training programs will attract and retain talent far more effectively than those that treat workforce change as a purely transactional adjustment.

The Rise of AI Agents

Another significant development shaping the near future of enterprise AI is the emergence of autonomous or semi-autonomous AI agents. Surveys suggest that more than eighty percent of executives plan to adopt AI agents within the next few years. These systems can perform sequences of tasks independently, coordinate workflows, and interact with digital systems in ways that resemble human collaboration.

However, the rapid expansion of agentic AI also introduces new risks. Organizations often experiment quickly with these capabilities but fail to establish adequate oversight structures. Leaders who approach this technology typically begin by deploying AI agents within internal operational processes rather than exposing them immediately to customer-facing environments. Internal applications allow organizations to test reliability, establish monitoring mechanisms, and refine governance policies before expanding usage externally.

Robust oversight mechanisms have become essential in this environment. Effective implementations include logging systems that capture every decision an AI agent makes, audit trails that enable review of those decisions, and clear escalation pathways for handling exceptions. Organizations must also define precise boundaries around what decisions AI agents can make independently, and which actions require human approval.

The Experience Economy as a Counterbalance

While artificial intelligence continues to accelerate digital capabilities, another trend is emerging simultaneously. As AI-generated content becomes increasingly abundant and inexpensive to produce, human experiences themselves grow valuable. The ability to gather in person, collaborate creatively, and share moments of meaning emerges as a powerful differentiator in a world where digital interactions are generated instantly.

This dynamic highlights an important strategic balance. AI excels at optimizing efficiency and producing digital experiences at scale. Yet the more frictionless and automated these interactions become, the more people seek authentic human engagement. Leaders must therefore design organizations that use AI to remove unnecessary friction while preserving opportunities for meaningful connection.

The Leadership Imperative

Despite rapid technological advances in AI, leadership capability remains the most important determinant of success. Surveys indicate that only about thirty percent of senior executives feel fully confident in their organization's ability to execute large-scale change initiatives effectively. Yet companies that invest in structured change management programs consistently outperform peers in both revenue growth and cost savings.

Successful leaders recognize that transformation requires sustained communication and visible commitment. They articulate why change is necessary, model the behaviors they expect from their teams, and allocate time for employees to learn new systems without penalizing early mistakes. They also celebrate milestones that reinforce the desired cultural shift, recognizing individuals and teams who successfully adopt new ways of working.

The Pragmatism of 2026

If the market learned in 2025 that generative AI is not a magic solution capable of transforming organizations overnight, 2026 will reinforce a different lesson: meaningful AI transformation is demanding work. It requires disciplined execution, strong governance, thoughtful workforce development, and continuous experimentation.

Organizations that succeed will be those that focus on clearly defined problems, build robust data and operational foundations before scaling technology, invest deeply in employee readiness, and balance efficiency improvements with the preservation of trust. They will learn quickly, adjust their strategies based on evidence, and resist the temptation to pursue innovation without purpose.

The future will not belong to organizations that moved first into AI experimentation. It will belong to those who moved most effectively, combining technological capability with operational discipline, cultural alignment, and a relentless focus on delivering real value.

21. THE FUTURE BELONGS TO LEADERS WHO CARE

The Organization That Ate Itself

I have seen technology deployments go wrong in many ways over my career. Broken systems. Missed deadlines. Cost overruns. Those are painful, but they are recoverable. What I am about to describe is different. It is the story of what happens when an organization deploys generative AI without a transformation plan and how the damage is not measured in failed software, but in broken people.

The organization was large, well-funded, and under pressure to modernize. Senior leadership had committed publicly to a Gen AI-powered customer experience. The mandate came down fast: integrate AI across the operation, reduce headcount, and show the board that the future has arrived. There was energy in the announcement. There was a budget behind it. What there was not—what nobody paused long enough to build—was a plan.

Not a slide deck. Not a vendor selection. Not a pilot roadmap. A transformation plan, the kind that answers the questions employees are silently asking the moment they hear the word "AI": What happens to my role? What does this mean for my team? How do I fit into what comes next? Does this company still need me? Nobody answered those questions. So, the organization answered them on its own. The unraveling did not happen all at once. It started with whispers.

Within weeks of the announcement, the informal networks that hold any large organization together began carrying a different kind of traffic. People were not sharing best practices or collaborating across teams. They were trading rumors. Who was getting cut? Which functions were being automated? Which leaders had information and which were in the dark? The rumor economy became the dominant communication channel—not because employees were irresponsible, but because the formal communication channel had given them nothing to hold onto.

Then came the self-preservation behaviors. When people feel threatened and uninformed, they do not rally together. They protect themselves. Teams that had previously collaborated began hoarding information. Managers who had been transparent with their direct reports started hedging because they did not know what to say and feared saying the wrong thing. Some employees began quietly positioning themselves for external roles, updating resumes, taking calls from recruiters during lunch breaks, mentally checking out of an organization that seemed to have already checked out on them.

The quiet quitting was pervasive, but it was the loud dysfunction that did real damage. People started undermining each other. Not in dramatic, confrontational ways, but in the subtle, corrosive ways that are almost impossible to address through normal management channels. Information was withheld in meetings. Credit was claimed for others' work. Blame was redirected. Small acts of sabotage disguised as miscommunication. When survival becomes the operating principle, collaboration becomes a liability. People who had been good colleagues for years began behaving in ways they would not have recognized in themselves six months earlier.

Morale did not decline. It collapsed. Engagement scores—for

those who still bothered to respond to the surveys—fell off a cliff. Several experienced team members went on medical or stress leave. These were not people who lacked resilience. They were people who had given years of tireless work to the organization and were now watching it disintegrate around them, with no one in leadership acknowledging what was happening.

The part that still troubles me: while all this was happening on the ground, the organization's leader was telling a completely different story upstairs. In C-suite briefings and board updates, the narrative was triumphant. The Gen AI launch was on track. Automation was scaling. The future was arriving on schedule. Slides were polished. Talking points were rehearsed. The word "transformation" appeared on every page.

Meanwhile, the customer experience numbers were telling the truth that the leadership narrative would not. Satisfaction scores were declining. Repeat contacts were rising. Handle times were increasing—not because of AI adoption friction, but because the remaining employees were demoralized, undertrained, and unsupported. The people who were supposed to deliver the customer experience were drowning, while the person responsible for their well-being was upstairs presenting a success story.

There is a particular kind of organizational betrayal that happens when leaders narrate victory while their teams experience defeat. It is not just dishonest. It is disorienting. Employees can see the gap between what is being said and what is being lived, and that gap destroys whatever remaining trust might have held things together.

People stopped wanting to be part of the organization. Not just the ones leaving, but the ones staying. The voluntary attrition was painful, and people who remained stayed because they had no immediate alternative. They sat in meetings they no longer believed in, executing plans they knew were disconnected from reality. An organization full of people who did not want to be there is not an organization. It is a waiting room.

Almost two years later, leadership finally recognized what everyone on the ground had known from the beginning: the organization needed a transformation plan. They hired someone to lead it.

By then, the damage was immense. The best talent had already left. The remaining team was exhausted, skeptical, and deeply distrustful of any new initiative. Institutional knowledge had walked out the door. Customer metrics had deteriorated to a point that would take quarters to recover.

The new transformation leader inherited not a blank canvas, but a battlefield—one littered with broken relationships, abandoned processes, and a workforce that had learned, through painful experience, that leadership's words and leadership's actions were two entirely different things.

Roles were eliminated during this period. In any transformation, that is sometimes necessary. Automation genuinely does change the nature and volume of certain work, and honest leaders acknowledge this rather than pretending otherwise. The way those eliminations were handled compounded the damage rather than beginning to heal it.

People who had worked tirelessly for years, who had carried the organization through its most difficult stretch, who had shown up every day despite the chaos, were let go in ways that lacked basic dignity.

Communication was impersonal. Timelines were abrupt. There was no meaningful recognition of contribution, no thoughtful transition support, no honest acknowledgment that the organization's failure to plan had made their roles casualties of leadership's neglect rather than inevitable consequences of progress.

When you eliminate roles without dignity, you do not just lose the people who leave; you also lose the dignity of those who remain. You lose the trust of every person who stays. They watched how their colleagues were treated. They drew conclusions about how they would be treated when their turn came. Customers adjusted their loyalty accordingly.

A Different Way: What One Leader Got Right

There is a temptation, in telling a story like that, to suggest that the alternative is simply not to change. That is not the lesson. Organizations must evolve. Roles will shift. Some positions will be eliminated as AI reshapes how work is done. The question is not whether transformation happens. The question is whether leaders treat transformation as something they do to people or something they do with them.

One of the most instructive examples I have encountered came from someone close to me who worked at Block, the financial technology company led by Jack Dorsey.

She said that Dorsey saw the same AI disruption that every CEO saw. He understood that generative AI would fundamentally change how work was performed across his company. He knew that roles would evolve and that some would be eliminated. He was not naïve about economics, nor sentimental about avoiding hard decisions.

But he did something that most leaders did not. He gave people time and agency. Dorsey issued a mandate that was deceptively simple: no meetings on Tuesdays. For several months, every Tuesday was cleared organization-wide and was designated for one purpose: reskilling. Employees were given an entire day each week, protected from the usual pull of operational demands, to learn the tools that were reshaping their industry to explore generative AI platforms, build new capabilities, and prepare for what comes next, not in a vague or aspirational sense, but through a structured, supported, and practical approach.

Think about what that communicates. Not in a memo. Not in a town hall. In action. It says: "*The future is coming, and I respect you enough to help you get ready for it. I am not going to automate your role on Monday and wish you luck on Tuesday. I am going to give you the time, the tools, and the trust to evolve alongside this technology, and if the day comes when your role here changes, you will be more capable than when you arrived.*"

Dorsey paired this with a broader philosophy of work that was equally forward-looking. Block embraced remote work not as a pandemic accommodation, but as a permanent operating model. The message was consistent: we trust you to manage your time, your development, and your contribution. We are building a company for how work functions now, not for how it functioned a decade ago.

When layoffs did come, as they have across the technology sector, the way Block handled them stood in stark contrast to what I described in the first half of this chapter. Employees who departed received generous severance packages. They kept their devices, the laptops and tools they had been working with, so their transition to the next opportunity was not disrupted by the logistics of returning equipment and losing access to their own work. It was a small thing in the scheme of a layoff, but it communicated something large: we still see you as a valued professional, not as a line item we are closing out.

The employees who had taken those Tuesdays seriously, invested in learning AI tools, and utilized the protected time to build new skills were meaningfully better positioned for their next chapter. They were not just told that reskilling mattered. They were given the time and space to do it. When the transition came, it was not a cliff. It was a step.

The employees who stayed were equally affected by how their departing colleagues were treated. They saw dignity and generosity. They saw a leader who prepared his people for the future rather than surprising them with it. This preparation shaped how they showed up the next Monday and every Monday after. They did not stay out of fear or inertia. They remained because they trusted that the organization's values were genuine and would be extended to them when it mattered.

Two Models, One Choice

These two stories represent the defining fork in the road for every organization deploying generative AI.

In the first model, AI is treated as a technology deployment. The focus is on automation rates, cost reduction, and executive storytelling. People are an input to be optimized. Change management is an afterthought. Communication is performative. The human cost becomes undeniable; leadership's response is to hire someone to clean up the mess—a year too late, with half the talent already gone.

In the second model, AI is treated as a human transformation enabled by technology. The focus is on preparing people—honestly, practically, and with respect—for a future that will look different from today. Communication in this style of transformation is authentic.

Reskilling is more than a slide in a deck; it becomes a day on the calendar. When hard decisions are made, they should be made with the kind of dignity that preserves trust—both for the people who leave and the people who stay.

The financial benefits of the second model over the first model are not even close. The first model appears to save money in the short term, but the costs of attrition, disengagement, degraded customer experience, rebuild time, and reputational damage far exceed the efficiency gains. The second model requires upfront investment in people, but it produces an organization that can execute its transformation because the workforce believes in it.

The Questions Every Leader Should Ask

If you are leading a Gen AI transformation—or about to—here are the questions that separate the two models:

Do your employees know what the plan is? Not the technology plan. If you cannot articulate what roles will look like in twelve months, how people will be supported through the transition, and what success looks like for the humans in the equation, not just the systems, you do not have a transformation plan. You have a deployment plan. Those are not the same thing.

Are you giving people time to evolve, or just expecting them to? Reskilling does not happen in the margins. It does not happen in a one-hour webinar or a voluntary evening workshop. It happens when leaders make reskilling a structural priority, protecting time, providing resources, and signaling through their own behavior that learning is not optional; it is the job.

Is the story you are telling upstairs the same story your team is living downstairs? Narrative gaps are trust killers. If your board presentation describes a thriving transformation and your employees describe chaos, you are not leading. You are performing, and your team knows the difference.

When roles are eliminated, how will you handle them? Before it happens, decide. What does generous look like? What does respectful look like? What does a transition that preserves someone's professional dignity look like? The moment you are in the middle of a reduction in force is too late to figure out your values. Your values need to be built before the pressure arrives.

What will the people who stay remember about how you treated the people who left? This is the question that most leaders forget, and it may be the most consequential of all. Every layoff is witnessed by the survivors. They are watching. They are taking notes. Their willingness to trust you, to go the extra mile, to believe in what comes next; all of it is shaped by what they see you do when the stakes are real and the cameras are off.

The Trust Algorithm in Practice

When I designed the Trust Algorithm, I built it for customers. But the truth I have learned over decades of leading large organizations is that the algorithm applies internally with equal force.

Trust Signals are not just what customers say. They are what employees experience in every town hall, every restructuring announcement, every decision about how departing colleagues are treated. Employees are constantly reading your signals, and they are far more perceptive than most leaders realize.

The Empathy Engine is not just for customer interactions. It is for the employee who is afraid that their twenty-year career is about to end because of technology they do not understand. Meeting that fear with a mandate and a deadline is not transformation. Meeting it with time, training, and honesty is.

The Recovery System is not just for service failures. It is for organizational failures, the moment when leadership recognizes that the plan is broken, that the team is suffering, that the narrative and the reality have diverged. The organizations that recover from those moments are the ones where leaders have the courage to say, publicly, "We got this wrong. Here is what we are doing differently, starting today."

The organization in the first story never activated its Recovery System. It let the damage compound for a year before acknowledging it needed help. By then recovery was not a correction, it was a reconstruction.

Jack Dorsey did something more elegant and more difficult. He built a "customer trust" algorithm into the operating model before the crisis by creating a culture in which people were invested, prepared for change, and treated with dignity when hard decisions arrived. He did not need a recovery system because he had built a system that did not break trust in the first place.

The future of AI in the enterprise will not be determined by the quality of the models. It will be determined by the quality of the leadership. Every transformation has a human cost. The question is whether you pay it in broken trust and damaged careers or in the investment required to do it right.

The organizations that win will not be the ones that move fastest.

They will be the ones that moved with the most integrity. We stand at a pivotal moment in the evolution of customer experience. For decades, organizations have pursued incremental improvements by optimizing processes, digitizing service channels, and introducing automation to increase efficiency. Today, generative AI offers something fundamentally different. For the first time, organizations can eliminate friction at scale, personalize experiences in real time, and respond to customers with a level of speed and contextual understanding that was previously impossible. Technology has matured rapidly. What remains unresolved is not whether it can transform customer experience, but how leaders will choose to guide that transformation.

At this crossroads, organizations are beginning to follow one of two distinct paths. The first path views generative AI primarily as a cost-reduction instrument. In this model, automation becomes synonymous with deflection—reducing human interactions, minimizing headcount, and measuring success through efficiency metrics such as contacts avoided or operational costs reduced. In the short term, this approach can produce impressive financial results. Budgets shrink, productivity metrics improve, and executives point to automation as evidence of progress. Yet over time, the limitations of this strategy become visible.

Customers quickly recognize the difference between an experience designed to serve them and one designed to process them. Conversations feel transactional rather than supportive. Complex situations become harder to resolve. Trust erodes quietly, often before leaders realize the shift has occurred. Customer retention weakens, churn rates increase, and the brand's reputation becomes more fragile amid competition. The organization may have optimized efficiency, but at the expense of the very relationships that sustain long-term growth.

The second path approaches generative AI from a fundamentally different perspective. Instead of treating technology solely as a cost-reduction mechanism, leaders view it as a **trust amplifier**—a tool that can strengthen relationships among organizations, customers, and employees. In this model, AI removes friction from interactions, improves the accuracy and speed of resolutions, and empowers employees with the insight needed to respond thoughtfully in moments that matter.

Rather than focusing solely on automating tasks, organizations redesign entire workflows to address the root causes of customer problems. Leaders who choose this path recognize that AI's full potential emerges only when technological innovation is accompanied by cultural and organizational transformation. They invest in change management alongside new systems. They measure customer effort, confidence, and trust in addition to traditional efficiency metrics. Most importantly, they deliberately protect opportunities for meaningful human connection, where empathy and judgment remain irreplaceable.

The financial results of this approach often develop more gradually. Trust does not compound overnight. But when it does take hold, its effects are powerful. Customers who feel respected and understood are more likely to remain loyal, increase their spending, and recommend the organization to others. They are also more forgiving when mistakes occur because the relationship has been built on a foundation of mutual respect. What begins as a customer experience strategy becomes a growth strategy.

Evidence from recent years reinforces the strength of this approach. Organizations that pair AI adoption with strong change management programs consistently outperform their peers in both financial and operational outcomes. They experience higher revenue growth, achieve greater cost savings over time, and report stronger employee engagement. The small percentage of companies that have already generated measurable profit-and-loss impact from generative AI share common characteristics: they focus on reducing customer friction, embed AI deeply into end-to-end workflows, and treat trust and engagement as leading indicators of financial performance rather than intangible ideals.

For leaders, this moment presents a deceptively simple question with far-reaching implications: should we optimize for the next quarter or build for the next decade? The attraction of short-term efficiency is understandable. Cost reductions appear quickly on financial statements, and the benefits are easy to quantify. Building trust, by contrast, requires patience. Its impact unfolds gradually and is more difficult to measure in traditional accounting terms.

Organizations can always extract additional efficiency from a process, but trust, once lost, requires years to rebuild. Leaders who prioritize trust-centered transformation often face more complex decisions and longer investment horizons. However, they build institutions capable of enduring beyond any single technological cycle.

Customers rarely remember the technology behind an interaction. What they remember is the experience itself: whether their time was respected, their concerns acknowledged, and their problems resolved with care. The algorithm that enabled the interaction fades into the background; the emotional impression it created does not.

In recent years, many organizations have struggled with declining trust among both employees and customers. Customers often feel processed by systems that lack a genuine understanding of their needs. Employees feel demoralized when new technologies are introduced without sufficient training or support. Leaders celebrate cost reductions while the underlying quality of customer relationships deteriorates.

The difference between organizations that recover from this erosion and those that deepen it is not the technology they deploy. Many of them use similar tools and platforms. The difference lies in leadership and how leaders choose to measure success, where they invest resources, and how they communicate their vision for the future.

Every leader leaves a legacy shaped by these decisions. Some will be remembered for implementing advanced technologies and achieving operational efficiencies. They introduced AI, optimized systems, and delivered measurable productivity gains. Yet those accomplishments alone rarely define lasting leadership impact.

Other leaders will be remembered for something more enduring. They built organizations where customers felt genuinely valued, where employees felt empowered to grow alongside new technologies, and where shareholders benefited from sustainable growth rooted in trust. These leaders used AI not merely to improve operations but to elevate the human experience at the center of their businesses.

Generative AI offers a once-in-a-generation opportunity to reimagine customer experience. Its true power lies not simply in its computational capabilities but in its ability to augment human understanding and connection on a large scale. Leaders who recognize both the technological and human dimensions of this shift will shape the future of their industries.

The challenge is therefore not simply to implement AI, but to lead with it thoughtfully. Leaders must decide how to align technology with their values, empower employees to use it responsibly, and ensure that every interaction strengthens the organization's relationship with the people it serves.

Technology will continue to evolve. Metrics will change. Strategies will adapt to new competitive pressures. The relationships built with customers, the opportunities created for employees, and the trust cultivated over time will endure far longer than any technological trend. The future of customer experience will not be defined solely by machine intelligence. It will be defined by the wisdom and courage of leaders who understand how to combine that intelligence with empathy, integrity, and purpose.

The question is not whether generative AI will transform customer experience; it already has. The real question, the one that will define your legacy, is whether you will build the trust this moment demands or spend the rest of your career explaining why you didn't.

Trust is not a feature you ship. It is a debt you pay, one human decision at a time. Every deployment. Every layoff. Every silence where leadership should be. You are writing an algorithm right now. Not the one your engineers are shipping, but the one your legacy will run on.

People will remember who you were when the machines got smart. Your customers will remember whether you saw them or saw through them. Somewhere, a young leader is watching you decide, learning what leadership looks like in the age of AI, and building their algorithm from yours.

So, write it like it matters. Because it does. Because they do.

PART V: APPENDIX

KEY STATISTICS AND RESEARCH SOURCES

Gen AI Adoption and Impact:

- 89% of organizations regularly use AI (McKinsey, 2025)
- 95% are seeing little to no P&L impact from Gen AI investments (MIT, 2025)
- Companies average 4.3 AI pilots, with 21% reaching production scale (BCG, 2025)
- 74% of companies struggle to scale AI value (BCG, 2025)

ROI and Investment Pressure:

- $30 — 40 billion in enterprise Gen AI investment (Various sources, 2025)
- 61% feel increased pressure to prove AI ROI vs. a year ago (Kyndryl, 2025)
- 53% of investors expect positive ROI in 6 months or less (Teneo, 2025)

Regional Adoption:

- Asia-Pacific: 45% Gen AI adoption at mid-to-high maturity (BCG, 2025)
- North America: 43% of the $590B digital transformation market (multiple sources, 2025)
- Europe: 45-70% behind the U.S. in Gen AI usage (McKinsey, 2025)

Transformation Success Rates:

- 70% of digital transformations fail to meet goals (Various consulting studies, 2025)
- 30% succeed when strategy, technology, and change management are aligned (McKinsey, 2025)
- 40% more cost savings for organizations with strong change management (Accenture/WEF, 2025)

Workforce Impact:

- 170 million new roles created, 92 million displaced = 78 million net increase 2025-2030 (World Economic Forum, 2025)
- 50% of employers plan to reorient their business due to AI (WEF, 2025)
- 66% plan to hire AI-skilled talent; 40% anticipate workforce reduction where AI automates (WEF, 2025)

Employee Adoption:

- 62% of 35–44-year-olds report high AI expertise vs. 22% of boomers 65+ (McKinsey, 2025)
- 48% would use Gen AI more with formal training (McKinsey, 2025)
- 45% would use more if integrated into daily workflows (McKinsey, 2025)
- Employees use Gen AI 3x more than leaders realize (McKinsey, 2025)

Change Management:

- 91% cite cultural challenges as an impediment to becoming data-driven; only 9% cite technology (New Vantage Partners, 2025)
- Only 30% of C-Suite are confident in their ability to drive successful change (Accenture, 2025)
- Only 25% believe teams are prepared to embrace change (Accenture, 2025)

Industry-Specific:

- 75%+ of banks investing more in data/cloud for Gen AI (Deloitte, 2025)

- 80%+ of manufacturers using AI for supply chain optimization (Industry reports, 2025)
- 80%+ of businesses globally plan to use AI chatbots (Customer service studies, 2025)
- 82% of executives plan to adopt AI agents within 1 – 3 years (WEF, 2025)
- 71% of consumers want personalization; only 43% feel they receive it (Deloitte/Think with Google, 2025)

www.ingramcontent.com/pod-product-compliance
Lightning Source LLC
LaVergne TN
LVHW052320200926
848481LV00002B/612